Bitten by the Bull Bug

Lennie Wallace

First published in 1997 by Central Queensland University Press

Second published in 2012 by Boolarong Press, Salisbury, Brisbane, Australia.

National Library of Australia Cataloguing-in-Publication entry

Author:	Wallace, Lennie
Title:	Bitten by the bull bug / Lennie Wallace.
ISBN:	9781921920578 (pbk.)
Subjects:	Wallace, Lennie.
	Ranch life--Queensland--Cape York Peninsula.
	Ranchers--Queensland--Cape York Peninsula--Biography.
	Women ranchers--Queensland--Cape York Peninsula--Biography.
Dewey Number:	636.01092

Designed and typeset by Carlin Yarrow in Garamond Narrow and Gaze Normal

Cover painting and all other illustrations by Harry Bruce

Printed and bound by Watson Ferguson & Company, Salisbury, Brisbane, Australia.

Contents

Acknowledgements

I wish to thank the family and my three Girl Fridays

Janice, Marie and Alice

who contributed more than they realise.

1

The March of Progress

By the advent of the 1960s I was well and truly a Peninsularite. After spending the first twenty years of my life as a nomad I'd found my real home. As a seventeen year old I'd gone to Cooktown with my older sister and our parents. For my father and mother it was their second stay in Cooktown where my father was firstly the Mining Warden's clerk and later the Mining Warden.

They'd been there some time before my sister and I joined them. She'd qualified as a pharmacist and I'd just matriculated at Brisbane Girls Grammar. Our arrival, with a young friend, greatly increased the female population of the township and caused quite a bit of excitement among the town's many bachelors. The warden's daughters were especially commented upon but it wasn't until I'd been made one of the mob that the verdict was made known to me—'one isn't a bad sort but the other's 'ighly educated'. We still haven't worked out which was which.

We both left again, Jean to travel the world and I, after an unforgettable holiday in the mustering camp at Butcher Hill station with my friend and fellow horse-crank, Ruth, to go nursing. Reluctantly.

In due course I returned, for it was decided that unless Ruth, who'd been hospitalised with a broken leg, had a permanent female companion of her own age, she was to be consigned to the nurses' home, also. Her pleading letter came at the right time. My father died and life seemed colourless. Here was a second chance. There was nothing I'd rather do than join her and the other ringers in the station camp. To be truthful, saving her from her proposed fate was but a secondary consideration.

The life at Butcher Hill, then only a day's truck ride from Cooktown, was as entrancing as I'd remembered it. I'd gone back there for my annual holidays but this time there was also an added attraction. Ruth's girl-shy and elusive brother, Bill, was home. We were married, after he'd written to get my mother's consent, just before my twenty-first birthday.

For years, one of the old-timers who frequented the bar at the Laura pub struggled to ask me a question, only to be angrily shushed by his mates before he had the chance to speak. Eventually they all faded away and he, the sole survivor, was able to pop his question: ' Who did the askin'?'

And I was able to tell him quite candidly, 'No one'. The decision was mutual.

As time went by we thought it was opportune to strike out on our own though Bill continued to be his father's right-hand man for many years. The Boss remembered a small block he'd bought near the East Normanby river. It was no further use to him. We could have it for what it'd cost him in rates. That sounded a fair enough deal and we accepted.

He'd pulled the house down using its beautiful pit-sawn timber in buildings at Butcher Hill and Lakefield. The cattle and horses acquired in the deal had long been absorbed into the Butcher Hill herds. The separate kitchen, connected to the main house by a covered walkway, was left to be used as a traveller's camp. A handy thing if the river rose too high to cross. Unfortunately the little building was destroyed by fire.

We were left with some fences in disrepair, a few charred stumps and a three-grave cemetery. Until we acquired the sharply-pointed iron that once formed the Cooktown convent fence and used it with bush-timber posts to build our first home, we lived under a low-slung tentfly. Neighbours, friends and relations proved kind and we soon scrounged the wherewithal to furnish it.

Until the road came, purchases were limited either to what could be brought in very expensively by 'plane or a little less expensively on the weekly launch to Cooktown and brought out — somehow — from there. A much larger vessel, the Wandana went on to Thursday Island every month and also carried freight for Cooktown.

For station owners these limitations made the purchase of even a few coils of barbed wire for a fence or similar station equipment a major and costly operation. Space was very limited and useless paraphernalia like furniture was relegated to the bottom of the list. A new wood stove might stand a chance but non-essential stuff like lounge suites and china cabinets never got a listing. What couldn't be made with whatever material was at hand could be done without.

My mother-in-law was a jack of all trades. She was an excellent horsewoman with a long string of pre-marriage show successes on the coast and the Tableland. She could crochet and dressmake beautifully, sew men's clothes like a professional tailor, cook and garden. A pianist with a good ear and superb sense of rhythm she made music to entertain herself and others, playing regularly for dances in Julatten almost up to her death in her late eighties.

Added to this she was a bush carpenter. The saws, hammers and chisels at Butcher Hill were 'hers'. When she lived in the 'old' house, built of slabs and complete with loopholes for rifles, she often shifted the internal partitions to enlarge or modify the rooms. With the 'new' house, one of the first 'kit' homes bought ready-to-erect from the old Queensland Pastoral Supplies in Brisbane in the 1930s, she contented herself with putting up the odd shelf and adding to the comforts of the house when the need arose or the Muse smote her.

I'd seen packing-case furniture and admired a three-kero-case bedroom cupboard with cretonne curtains and a horsehair-padded cretonne seat that she'd made. Kerosene cases were going out of use fast with the advent of the larger drums required to keep modern kerosene fridges operating, but the Main Roads introduced a substitute — gelignite boxes. They were a little smaller than the kero case but were joined with eye-catching, very neat and rather small, dove-tailed joints.

When the need arose I decided to acquire a few of these to make a nursery cupboard. I used a bit of bush-barter. At smoko, when the Main Roads called, they got a fruit cake and fresh eggs to take back to camp. In return, I became the proud owner of three sound boxes and their tops. With them came suggestions. Why not be a little more ambitious? The dove-tailed boxes would make ideal drawers. Why not make a chest of drawers for the baby clothes rather than curtained shelves? The idea grew.

The best part of a sheet of almost-new ply was donated to cover a scrap-pine carcase. With gelignite box-lid runners, gelly box drawers also fronted with ply, painted pale blue and fitted with store-bought drawer pulls and decals of gamboling lambs (brought from town per courtesy of the Main Roads truck) the effect was very pleasing. I wondered why no one else had thought of such a marvellous use for the boxes.

A couple of months later when I took out the 'town' clothes, little smocked shirts and dresses, from the bottom drawer, I found the reason why. Something, presumably in the composition of the gelignite, had turned the clothes a brilliant, jaundiced yellow. In the other drawers, everyday clothes and nappies, being used and washed more frequently, were just a creamy off-white that I'd attributed to the slightly muddy creek water.

No manner of soaking and scrubbing removed the gelly's influence. It permeated the absorbent pine permanently and proved a long-lasting clothes dye. So much for pride and dove-tailed gelignite boxes.

Three children were born while we were at Harvest Home. John and Nancy were both Cooktown babies. Billy Kid saw the light of day in Atherton. There was no doctor in Cooktown and all but emergency maternity cases were tabu.

I was in a bit of a fix. Not having any relations handy to act as baby sitters and having to go to Cairns to pick up our first little green Land Rover, I packed the hospital ports and took them with us. Just in case.

And just as well. I called to see an old Grammar School mate of mine who immediately solved the problem. She lived on the outskirts of Atherton. I could stay with her. Nancy would be company for her two boys and Bill and John could go home with the Land Rover.

I didn't even make it to the first antenatal clinic day but had young Bill instead and was out before I should have been in.

Then our home address changed again. Bill was offered a working share in his old love, Lakefield, and was seriously considering the proposition until his father made a counter offer. We could have all the Butcher Hill country below the Block Fence. This was a short fence stretching from hill to hill at the Laura end of the run that 'blocked' the cattle from going down along the Laura River and the drovers' road where they could be led astray by ill-mannered droving drop-outs. We could also have the hut, yards and paddocks on the homestead side of the Block Fence.

Bill took up the offer. Our new home acquired the name Crocodile after a big waterhole of that name down the river and we moved in. I put in a tender for forty pounds ($80) for an old building behind the Post master's residence in Cooktown which we pulled down and re-erected.

With the money I'd made free-lancing for various publications I bought a two-way radio to use in conjunction with the Aerial Ambulance Base in Cairns. The Base, VKA, held scheds during the day during which weather reports were handed in and telegrams sent or delivered. Emergency calls could be made at any time and if the Ambulance bearers couldn't deal with the case themselves (with the help of a standard Flying Doctor-type medicine chest) they called on medical advice. If necessary, an evacuation by light plane would be set up taking the patient to the Cairns Base Hospital.

The Ambulance bearers were, almost all, men of varied talents. One in particular, Charlie Harriman, was as multi-skilled as anyone could hope to be. Charlie kept us and our families in good working order, treated our animals, resurrected seemingly-dead engines — the lot, and all by radio. Whether he was actually as knowledgeable as he seemed no one chose to doubt. His advice worked ninety-nine point nine per cent of the time and was a great boost to anyone's flagging self-confidence.

In one instance my radio was doing weird things. The rivers were up, the road out, so there was no way to send it to Cairns for repairs as one bearer suggested. It was the time of year when we really needed our transceiver.

Charlie to the rescue. He told me to unscrew the back of the radio and with a bit of questioning gave both his diagnosis and treatment. The latter included the use of a sharpened matchstick inserted in one place and the folded cardboard from a Log Cabin tobacco packet in another. Thus treated, the radio never faltered.

Years later, when we had a phone connected, I gave the transceiver to a friend suggesting she have it checked-out before she used it. The unit was operative and the matchstick and folded cardboard still in place. Reluctantly, the technician removed them, replacing them with something more up-market.

Once the road was sketchily surveyed, approximately the route taken west of the dividing range by wagons and stock from the southern settlements to Byerstown and Maytown on the Palmer goldfield in the 1870s, adventurous motorists set out to emulate the feat of the two Kiwis who went from Cairns to the Tip in a Baby Austin during the 1920s.

With all the steep pinches and river crossings no one met with any success until the Main Roads moved into the picture. Camps were set up in odd spots along the route selected. They didn't start at Mt. Carbine or Curraghmore or wherever the road ended and head progressively north. Instead the roadwork was haphazard, a formed section here and then miles of almost two-wheel-track stuff dodging the trees and monster antbeds until Head Office splurged again on a bulldozer for another fragment of cleared road. Bridge gangs, after several attempts, got bridges more or less permanently across the trickiest streams and moved on to other sites.

The locals found treasure troves in abandoned Main Road camps that brought out inherent bowerbird instincts. Bolts with or without nuts to fit, nails, billycans, useful lengths of sawn timber, pieces of steel rod, a handleless shovel or axe. Wonderful items of advanced technology. First on the scene had a windfall.

More lasting treasure-hunting occurred when, in the early 1950s, a camp was made on a snug little flat adjacent to a waterhole in the Laura River. This was despite warnings from Len Elmes, passing through with a mob of cattle, that the tangled limbs visibly wedged in high tree-forks and the corresponding scars on tree trunks were put there by previous floods.

As usually happens, the floodwater came down in the dark of night. Fortunately the men had no problem saving themselves and their personal gear. Trucks and machinery were also moved to higher ground. Some of the drums from the fuel dump were rescued but odd ones were found by musterers over the years, miles downstream and in the most unusual places. Every Wet for several seasons something new and interesting would be revealed as a fresh run of water shifted the sand, an enamel plate here, a useful length of chain there, a billycan, a shovel.

As the road became relatively permanent from 'after the Wet' until 'the storms' station owners invested in four-wheel-drives, a few indestructible ex-army jeeps and the larger 'blitzes', but mainly Land Rovers, the little green variety with canvas tops. Later the long wheel-based grey ones which would carry a fair load of station requirements came into their own.

Bush Pilots, a small airline set up to take over the mailruns and services from the packhorse mailmen, began landing on the valiant little strips, often cleared by hand with nothing more advanced than axes, picks, crowbars, a fire-stick and plenty of man- or woman-power. They were also used by the Aerial Ambulance based in Cairns for rescue missions. The strips were scattered through the Peninsula north of Laura.

The bauxite that coloured the Weipa cliffs a bright orangey-red became commercially desirable and the road and roadmakers pushed further north. Other miners came in by four-wheel-drive or helicopter and for a short period oil prospectors abounded.

Bill Raymond, returning to Kimba with a truckload of gear, camped at the top of the Byerstown Range. Just on daylight Bill started on the second stage of his journey and almost collided head-on with a B.M.R. helicopter pulled up in a creek-crossing while the prospector filled his sample bags.

One of the miners who made bow-waves in the deepening bulldust as he headed for the top of the Peninsula was a man who later made money out of southern sandmining. He liked the rich red basaltic country he saw through the bulldust in the Laura valley, comparing it to good farming land in other parts of the state. Some years later the miner, Clive Foyster, bought the land that had intrigued him so much and came very close to seeing his dream of a thriving farming community realised at Lakeland.

The road, extravagantly named the Mulligan Highway after James Venture Mulligan, the adventurous prospector who had a hand in most of the far northern gold strikes, was declared by someone in Brisbane to be 'all weather'. As a result the Cooktown to Laura railway with its weekly railmotor was doomed.

Despite loud local protest the line was closed, the rails pulled up, bridges destroyed and the rail carted away by semi-trailers. As it happened the bridges had mostly been repaired the previous financial year thus inflating the maintenance costs and providing what the southern experts regarded as a very good case for closure on economic grounds. Some of the line returned soon after it was taken away to serve as telephone poles taking the phone to Laura by a more direct route.

Fears that a road out would soon depopulate the North were unfounded. Apart from a few ex-railwaymen, no one left. Except for the Territory there was no other frontier to offer sanctuary. People moved in, some on transitory business or sight-seeing and others to stay. Big companies, American and Australian, bought several of the better and larger cattle runs. With progress raging, our uncomplicated lifestyle began to crumble.

Miles Morris of Maitland Downs lived in a small bachelors' hut at the top of the Byerstown Range. Every so often he collected what mail accumulated for him at Butcher Hill. In the 1940s, Miles, his brother Henry (Hank) and Ian Pratt came north from Central Queensland to take up Maitland, an abandoned 'vacant' run of Crown land that had once been one of a string of State-owned cattle stations in Labor's long reign much earlier in the century.

Making bow waves in the bulldust

Coming from civilisation Miles knew a little about it — both the pros and the cons. Mostly he saw the cons.

He had one of the first trucks in the area, a Morris like himself, a two-way radio and sundry portable pumps and engines. All very modern high-tech stuff for the Peninsula. One evening Miles drove down for his mail with his feathers properly ruffled. It was during the '60s drought and his normal source of domestic water had dried up. Putting a portable pump and a square ship's tank on the back of the truck he drove a mile or so to another spring near some old mango trees, right on the road.

As Miles guilelessly waited for the tank to fill, dreaming a little and thinking happy thoughts, a Land Rover drove up. On the door was the State emblem and the legend "Irrigation and Water Supply".

"Do you have a licence to pump from this stream?" one of the occupants called to Miles as they slowed to negotiate the parked truck in the roadway.

We tried to persuade Miles they were only joking but he wouldn't be comforted. He didn't have an overly-developed sense of humour at the best of times and this was definitely not an opportune time. Knowing civilisation and bureaucracy better than we did, he saw it as a decidedly bad omen.

Perhaps he was right.

2

Storks Are Waterbirds

While the practice of cows calving in the stressful dry weather was looked on with disapproval, the converse was true for expectant women. Valuable cows lost so much condition in the dry that the added stress of lactation could, and often did, cost them their lives.

On the other hand, if it rained, the rivers came up and the roads went 'out', an unsuspecting man could easily be left in a very uneasy situation. Men are reluctant to get flooded-in with women in labour. Bowie Gostelow had fearlessly guided babies into the world at Violetvale but few were of the same calibre as Bow. When Iris Wallace had Beth at Merluna, her husband Hardy made sure he had more important business miles away.

Our fourth baby was due in March, definitely not a good month for travel, though, unlike Iris and Vivienne Gostelow, I had a Land Rover and now, a road.

Bill was still droving but Hopevale Mission were to inspect a mob of horses out from Mareeba and wanted to bring them back with Bill's plant. This suited both sides. Bill came back early on his own and Archie Gibson and the other 'boys' stayed in Mareeba to attend to the horse venture.

Most expectant mothers went to town a fortnight 'early' and I was due to leave about a week after Bill came home. As he'd been away since New Year I was looking forward to his company before I went to Cooktown to wait.

Everything was under control. I had a young Daintree girl, Cissie Cobb, to help with the children and Mrs.Yeatman was coming to oversee John and Nancy's correspondence lessons. Mrs.Yeatman was a remarkable woman to say the least. Well-educated and a woman of refinement she ran not only the Post Office at Rossville for a record number of years but also her own tin mine. Very active in mind and body she was well into her eighties when she first came to Crocodile.

She loved being in the bush and helped many young mothers, providing companionship for them and education for their children. Almost ritually she returned to Crocodile each year in September or October when the Primary Correspondence School's Paper 32 was due. This was the all-important paper that decided whether the pupil advanced to the next class in the new school year or stayed in the same grade. One week was allowed for revision and then Paper 32 was done with the solemnity of examinations of far greater importance.

My ports were packed. Clothes for before, after and during 'hospital' and another suitcase whose lid had to be sat upon to close it, of snowy white nappies, tiny singlets and nighties.

Everything was under control until early on the morning after Bill arrived home when visitors drove in. John and Mick from the Main Roads camp and a very lame drover, Ted Youngman. Ted. bought a pair of riding boots in Laura and in the pouring rain and endless river-crossings the dye infected a small puncture probably made by a spear-grass seed on his instep.

His foot swelled like a footrot bullock's. The skin stretched so tight it was a wonder it didn't split and was so red and inflamed you could almost feel the heat and pain emanating from it. Riding, even with it cocked up over the saddle pommel in front of him, must have been agony.

"We knew you were going to take Lennie to Cooktown, "John said.

"But not just yet!" I felt like protesting.

"Ted thought if he went in too, Matron could have his foot fixed in time for him to meet his cattle again to go up the Byerstown."

Ted nodded his head in agreement. In a week his foot would be back to normal and he could be with his mob over the roughest part of the trip, the steep and stony Byerstown Range. Everyone had faith in Matron Brown's capabilities.

Faced with such logic I knew I couldn't beat 'em. In any case I was greatly outnumbered. I might just as well cut my losses and join 'em. It would be a lot simpler.

My bulky hospital ports were put in the back and we left in the little Land Rover after the traditional, fortifying cup of tea.

So much for wishful thinking.

Our first unscheduled stop was only a few minutes down the track — the unbridged Laura River at Carroll's Hole. Despite a plastic bag being wrapped around the distributor the usually unstoppable, go-anywhere Rover bailed-up midstream.

Immediately the doors were opened, water rushed through just under seat level. I crossed my fingers and hoped the nappies etc. were high and dry on the spare wheel lying on the back floor. They were.

John and Mick were still with us so all hands pushed while I steered until the bank was reached. Ted insisted on pushing too.

A dry rag was sought. None could be found so I ducked behind some bushes and obliged with my halfslip. Apparently the battery was wiped dry too as when I recovered my undergarment the battery acid had turned it into something resembling very fragile lace, more holes than fabric.

Although progress was slow our next stop proved to be the end of the road. Boggy Creek between Butcher Hill and Springvale was coming down in waves that would have gladdened a surfie's heart. A quick creek-bank conference and a change of plans. Butcher Hill had recently acquired an airstrip. We'd go back there and fly to Cooktown in style with the Aerial Ambulance. Mutual congratulations were exchanged on the so perceptive decision. We went to Butcher Hill, contacted VKA, the Cairns Base, and the plane landed.

I was a little averse to small planes but having no acceptable excuse I boarded the aircraft with Ted and his swag, the pilot, the Ambulance bearer and one suitcase. The 'baby' port could follow me to Cooktown later.

With much waving and cheery, confidence-building smiles from our farewell committee we became airborne and headed east.

At first I was interested in the scene below pointing out to Ted, whose foot was anything but comfortable, the West and East Normanby rivers and what I thought was the rooftop of old Harvest Home, our first 'station'.

Suddenly, without a breath of warning, we found ourselves enveloped in a fug of thick cloud. Visibility nil.

"I'll have to re-fuel in Cooktown," the pilot remarked conversationally to the Ambulance bearer as we tried to skirt the cloud. "Getting a bit low."

He called Cooktown on the radio several times but could get no answer. Nor was Cairns radio any more forthcoming.

We flew through bank after bank of cloud. Occasionally a mountain loomed menacingly. The flight was anything but smooth. I was healthy when I left home but I certainly didn't feel that way for long as the plane pitched, plunged and rolled. I thought of the pilot's request for more fuel but philosophically decided the quick ending might be preferable to the current death by a thousand knife-cuts.

Men are reluctant to get flooded-in with women in labour.

Ted sat hunched in his seat, obviously in pain, and said nothing.

"Ever been here before?" the bearer, veteran of dozens of flights like this, cheerfully asked the pilot.

"Not that I know of," replied the pilot equally cool and unflustered. "I might put out to sea. Should strike the coast around Bloomfield. Be smoother out there."

Visions flashed through my mind of the Ambulance plane that had, only a few years back, crashed into the sea with pilot, bearer and a broken-legged patient. The latter drowned. Ted exhaled a huge sigh that made me think he was tuned to the same thought-wave.

We hit the coast. Not at Bloomfield but at Port Douglas, much, much further south. Almost in suburban Cairns. The last bit of the trip was a breeze and, of course, we landed safely.

Very pleased to be on terra firma (and the firmer the better) I began to walk across the light-plane tarmac while the bearer gallantly carried his gear and mine. But try as I would I couldn't walk straight. This is, I thought, how a drunk must feel and lined-up the small terminal making a determined effort to steer straight for it. I still drifted forty-five degrees off course. It was a strange and rather disconcerting sensation.

Once assembled at the edge of the tarmac Ted and I reassessed our positions. He was stranded in Cairns with his swag — but without his droving plant and cattle — and I was there, all intact except for the nappies and my post-baby clothes. We decided we'd go to the Bellview on the Esplanade. The proprietess, Mrs.Gallogly, had the well-deserved reputation of looking after bushies — especially expectant mothers — extremely well. She placed the mothers in rooms as close as possible to her own and never looked at all put out when woken in the middle of the night, at her own request, to call a hospital-bound taxi.

However, another surprise was in store for me. The ambulance we were travelling in veered off the Esplanade and pulled up at the Maternity Hospital.

I protested loudly but unsuccessfully. I wasn't 'due' for three weeks!

"Tell that to Sister Gibbons," was the Ambulance man's heartless rejoinder.

I did but the good Sister decided I needed a 'rest'. She also discovered the spaying scissors Bill had given me to have sharpened in Cooktown. They're used to remove cows' ovaries so that they'll fatten rather than breed.

"Just what's needed around here!" said Sister brandishing them over her head as she swept out into the main ward.

Suddenly I found I couldn't keep awake and by the time Sister returned to put the scissors safely in my locker I was asleep.

Sister was adamantly opposed to my pleas to be discharged. A rest would do me no harm. I explained I had to make some final arrangements for Bill's droving plant to come back with the Mission — and get the spaying scissors sharpened.

Finally she relented. A little.

"All right, you can go down town in the morning to attend to your business but be back here for lunch. The last mother I let out, had to get a taxi home as soon as she got there. Her waters broke at the War Memorial." The Memorial was then in the very centre of town. That piece of news had the effect Sister Gibbons desired. I know when I'm beaten so accepted the compromise.

One of the first people I met was Ted, sitting with his leg propped up on a bench outside one of the hotels. He'd been to Out Patients and already his foot seemed greatly improved.

"Weren't you scared," he asked, "when the bearer asked the pilot had he been in amongst all those bloody mountains before?"

"I was too sick to be frightened," I told Ted, not entirely untruthfully.

"Well," spluttered Ted. "I was too bloody frightened to be sick!"

As Sister predicted, I made the best of my enforced stay and enjoyed the 'rest'. It took some time to get a message to Bill that I was in Cairns rather than Cooktown. His telegrams to me were re-directed but it took considerably longer to be re-united with my missing port which Bill's mother had very kindly sent in to Cooktown at the first available opportunity. Our second daughter, Laura, was born and I was allowed out to Bill's uncle, Arthur, who lived near the hospital and who had been my one regular visitor.

As is usually the case, the people (men) who were so anxious to be rid of an expectant mother weren't anywhere near as keen to get one (me) back home with the baby. The danger past, they could relax.

Not long before Laura's birth we purchased Louis Fischer's property on the Daintree River. Louis had been our benefactor for many years selling us top quality bulls and heifers on time-payment and more often than not including a special 'pet' heifer of his for me. Officially we took over on January 1st but on the last day of the old year Bill's father broke his leg when a cow cannoned into him during a bid to escape being yarded. He was riding a little grey mare, one of my ex-saddlehorses and also named Laura. The blow was hard enough to break one of Laura's ribs as well.

With his usual acute sense of organisation, the accident happened on the new airstrip and it was more or less a case of easing the patient out of the saddle, into the plane and off to Cairns Base Hospital.

Bill, who had been in Daintree, returned to finish the muster before heading off with his droving mob to Mareeba.

It was now the traditional time to put the Butcher Hill bullocks together for sale, so getting me and the new baby home was the last thing on the agenda. Nothing — but nothing — must hold up a bullock muster. A message came through to go to Daintree with Mac Edmonds, a cattle buyer heading that way, and to wait there. Patiently. Eventually, after I'd gone down to a bout of Daintree fever and the bullocks were safely away and no longer needed care and attention, I was picked up and taken home.

You'd think that any reasonably sane person would learn from experience but no, almost exactly two years later the stork was again flying low getting ready to land. Again

I was waiting for Bill to come home to take me to Daintree where a prospective buyer was to look at some bulls. That accomplished, he was to drive me to Uncle Arthur in Cairns. I'd been warned by Matron that doctorless Cooktown was out of bounds to loaded storks.

Again Bill just got back. This time from mustering with his brother-in-law George Ahlers. He'd left his plant horses in the paddock at the top of the range where Miles Morris used to live, to be collected later. George and Bill's sister Joyce were now the owners of Maitland Downs. Miles, Hank and Ian had moved further north.

Our old drovers' cook, John Barry, was at Crocodile doing up saddles and gear and making himself generally useful so when Bill heard George had to put a mob of sale cattle together urgently I decided I'd go along too to see if Bill was wanted for the muster. It wouldn't take long and I saw so very little of him in the first half of each year.

Laura came with us. We left young Billy with John. They got along well. Billy imitated John's limping walk to perfection — to John's fond amusement — and the pair limped everywhere in complete unison.

The two older children were at the now Uniting Church's hostel in Coen and attending the local school. Few of the Peninsular children did correspondence lessons so most of John and Nancy's mates were already there. With no Cissie to help Mrs. Yeatman — Cissie had married — this seemed the best alternative. Johnnie was nine, Nancy going on for eight.

All went well on our trip out. We'd arranged to meet Joyce and George at the Palmer River cafe recently established by Tom and Bub Edmonds. It was then on the Mareeba side of the river. The Palmer was lapping the log bridge when we got there so Bill prudently left the Land Rover on the northern bank and we walked across. It didn't take long to decide about the muster and we were about to return home when ominous black clouds appeared over the headwaters.

Laura and I were to stay, Bill to go home, get the necessary 'hospital' ports and return. Simple. It would have worked out beautifully if the electrical parts of the Land Rover's innards hadn't been wet by splattering puddles before Bill pulled up on the bank of the hubcap-deep Spear Creek — where Johnnie Douglas had drowned — to put plastic over the distributor. He was halfway across when the engine conked out and a dreaded wall of water appeared suddenly upstream.

Later, in the post-mortem, I inquired sweetly why didn't he try to crank it out by turning the engine over enough times with the crank handle to move the vehicle out of trouble. His answer was short and succinct.

"You can't crank with your head under water."

While we at the Palmer were blissfully unaware of the drama, Bill left the Land Rover at the bottom of Spear Creek and set off on foot for his horses in the Old Maitland paddock. With the water so turbulent and muddy it was hard to know just where the Land Rover was. Just below the crossing was a deep hole, almost permanent, and he hoped it hadn't been swept into that.

Bill had been riding a chestnut mare of mine, a bit of a pet and quiet to catch. With a change of luck he caught her easily and using a greenhide halter hanging in the old hut rode her bareback across country and over the range to where the Raymonds were camped on a fencing job at the Twelve Mile.

In next to no time their trusty old blitz truck was mobilised and driven to Spear Creek where the water had fallen almost as fast as it had risen and revealed the Land Rover thankfully not in the deep hole! It was rescued and taken back on the blitz to the Raymond's camp where its vital parts were cleaned of sand, mud and other extraneous matter and dried out in Bel Raymond's oven.

The battery, chock-a-block with sand and mud, was a write-off and for as long as we owned the old Rover whenever we hit a sudden bump, sand showered down from a secret place in the cab roof.

Gordon and Bill Raymond, both good mechanics, decided that with Bel and her oven's help, they could get the vehicle mobile. This was the stage reached when we got the news per two-way radio of what had happened. Plans were changed. My Bill, Gordon and Bill Raymond, using a Raymond battery and the dried-out parts, decided to go for a test-drive to pick me up. That way, I could pack my own ports. I also had Indian outfits made from chaff-bags and dyed chook-feathers to be packed for an important fancy dress event in Coen, where, I learnt later, the few costumes other than chaff-bag and feather Indians were cowboy regalia.

While the Edmonds escorted us to the Palmer bridge, Bill and the Raymonds guided Laura and me across. The water was strong, about knee-deep for adults and the logs which formed the decking were slippery so I was glad of a strong hand or two to hold. One of the men carried Laura. The Land Rover was left with the engine running. No sense in tempting Providence. The re-conditioning however was almost a hundred per cent complete. Apart from the not-too-remote chance that it mightn't start again it suffered some mysterious (to me) electrical fault that made it necessary to insert silver paper between the points while the ignition was off. Otherwise the battery would flatten and the motor consequently not start.

The condition of the road didn't allow for speeding but was more suited to the state of the motor. We drove slowly and carefully back to Carroll's Hole. Here we were to get out and walk home while the Raymonds took the Rover back for a bit of fine-tuning, to return in the early morning to meet us.

Despite my assurances that I could cross the river safely, the Raymonds gallantly insisted they wait to see us over to the other side. Wet T-shirt competitions had not yet become the rage and I felt that the picture I made as I reached the other side to wave goodbye to our rescuers, almost nine months pregnant and wet from the neck down, would be more 'revealing' than 'appealing'.

John and Billy Kid cheerfully got a meal ready while I hastily gathered nappies, waisted, open-fronted dresses etc. into one suitcase, pre-baby clothes in another and bows, arrows and Indian clothes in a parcel to be posted. Bill sorted-out horses and gear for the morning's start.

Bill 'killed' when he came home from mustering and, as the Raymond's were out of fresh meat, Bill and John filled a split-bag with beef to take to them. A split-bag is a cornsack sewn at the open end and 'split' through one thickness at the middle to allow things to be put into the pouches thus formed. The filled bag is then adjusted over the back of the saddle with the bulky contents just behind the rider's thighs.

John was to ride the horse with the split-bag. It wasn't a terribly successful decision as the horse had never felt a laden split-bag across his back before and might even have objected to the smell of the freshly-slaughtered beef. Added to this John had broken his neck several years previously and though he had taught himself to walk again against all expert predictions, was rather limited in the use of his legs.

Getting on a rather tall horse and lifting his leg over a bulging bag of beef was slightly over the limit. John got stuck. I was holding the horse's head as a precautionary measure to make it stand still but John could get neither up nor down. Bill, saddling Johnnie's little Timor pony for me, saw John's predicament and quickly lifted John's stiff leg up and over while he propped the rest of John up on his shoulder. We all expressed fervent hopes John wouldn't have to dismount before we arrived at the river.

Bill had his horse saddled with a cushion tied in front for Billy Kid and both suitcases piled high on a second saddlehorse making its debut, a trifle unwillingly, as a packhorse. The cases were too big to fit in the packbags and had to go on top of the packsaddle as top load. It towered so high I was sure the mare would turn-turtle and capsize. She didn't but her distorted appearance created terror when we approached her mates in the house-paddock. They fled ahead of us, snorting and bucking in mock alarm.

Bill held Pony's head while I clambered on. He was a very small pony and Johnnie's tiny saddle was the only one that fitted him. I wasn't commensurately tiny. I, too, had a cushion in front for Laura, but with my own 'front' rather prominent and Pony being typically short in the rein or length of neck, this put Laura halfway to his ears. Added to this she refused to be parted from a precious Teddy Bear and I had to stretch my arms out enough to encompass Teddy as well.

John laughed heartily at my predicament. Whenever Pony, as ponies like to do, put his head down for a tempting snatch of grass, I had no hope at all of getting his head up or his feet back in the right direction until he was ready. With his short neck sloping steeply downwards I was fully occupied trying to stop Laura (and Teddy) and myself sliding over his short, pony ears.

Bill rode off first leading the 'packmare'. John came up behind me to chivvy Pony on in case he decided to take a few too many bites of grass.

The Raymonds were waiting. We rode across the river instead of wading and Pony being so short in the legs I transferred to one of the other horses to make the crossing. Bel had done more to the drying-out process of the electrical parts, keeping her stove going all night. The motor ticked over well, though the points remained a problem. Gordon and Bill left us at their turn-off smiling and confident we'd have no more trouble.

Not far past there we struck some Main Roads men who appeared on foot from nowhere in particular having very successfully bogged both a truck and an endloader in the middle of the road.

Naturally, courteous and helpful Bill pulled up to assist. Combined effort got the loader out after an hour or so of concentrated labour. All hands took a moment off to roll a smoke before starting on the truck and one of the men asked Bill where he was heading.

"Taking my wife to hospital," he said.

Heads swung my way. The men were aghast. Until then I hadn't been anywhere near as interesting as the bogged machinery and hadn't merited even a momentary glance.

"You get going," they said assuring us that now the loader was free the truck would surely follow. Considering they had bogged the loader in an endeavour to free the truck that wasn't necessarily so but they were insistent.

Reluctantly Bill agreed and off we went, left both children at Maitland and continued on — not to Cairns and Uncle Arthur which in the circumstances would have been more prudent — but to Daintree and the prospective cattle-buyer. The market had been more down than up and Bill wanted me to be there should we have to drop our asking price to meet the market. So much for honouring business partnerships in times of trouble.

Of course, the night we arrived in Daintree, to a property on the 'wrong' side of an unbridged river, it began to rain. By bedtime it was clear there was very little hope of a buyer getting 'in' and not much more of me getting 'out' — by road.

Louis' s nephew, Wadgee, was at the farm. He was brightly reassuring.

"It won't last. It'll stop when the tide turns and follow the tide out to sea."

Surprisingly some Daintree rain does just that.

But not this lot. I wasn't convinced. People in Upper Daintree all have rowboats as a matter of course. They need them for human emergencies as well as to push cattle likely to be stranded by rising water onto higher ground. As the river rises, the boats are moved progressively to higher anchorages.

From the watery noises in the blackness outside and the short space of time Bill and Wadgee were away, I gathered the boat was now at the last position, tied to a steel fence post hammered-in for just that purpose a few yards downhill from the house.

It was a long night spent with pricked ears strained to interpret all kinds of unnerving noises outside.

At breakfast I nobly resisted the temptation to say, "We should've gone to Cairns first." I can't say the thought wasn't topmost in my mind.

The men looked very gloomy. Obviously deep in thought, their minds were in overdrive. They'd have to find a way out — or they'd be stuck with me.

The little rowing-boat was the answer. We piled in with Wadgee, a lifetime's experience behind him, at the oars. As the road on the other side of the river often went under water we'd left the Land Rover on a patch of high ground at a friend's place downstream.

We landed there, had the obligatory social cup of tea though we'd just breakfasted, took the suitcases aboard and set off again. With the two large ports there wasn't much room with the three of us but I was determined to enjoy the trip. I'd never have the chance again and after all, people travel half-way round the world to experience gondola rides in Venice.

It was quite pleasant. With tide and flood combining to make light work of the rowing we sped along swiftly, easily dodging floating logs and other impediments. Not a bump or a boggy spot in sight.

Once in the township, Wadgee phoned his fiancee who immediately drove up with Wadgee's brother from Mossman to collect me. At Bill's suggestion I rang the stock and station agent and told him the inspection was, for the moment, 'off'.

As the creeks, complete with crocs and freshwater sharks, were rising quickly between Daintree and Mossman we wasted no time getting out. But even before we arrived in Mossman the creeks rose behind us cutting off road access and because of the phone line's proximity to road and river, causing the phone to go out in sympathy.

Bill and Wadgee took turns to row back, not so merrily this time with tide and current against them and rafts of floating debris stretching from bank to bank for them to get 'bogged'in. Bill didn't enjoy the experience and their shoulder muscles were sore for days.

In Mossman things were considerably more snug and comforting. There was also a hospital where Wadgee's Irene nursed. I was in the best of hands. Irene decided she'd drive me to Cairns herself in the morning. She could do some neglected shopping while she was there.

That night, too, it rained and rained. At breakfast the radio said the coast road was 'out' due to a landslip. No worries. We'd go up the Rex Highway to Mareeba and back down through Kuranda to Cairns. A road report immediately following the news quickly squashed that alternative. The Collins Bridge on the Kuranda road was under water. The consistent drum of rain on the roof was beginning to have a distinctly vengeful tone to it. We sat there silently, wondering when — and if — the rain would ease when someone called from the front door.

It was a very young man from the stock agency in Cairns. He called in to get directions on how to find Bill to tell him the would-be buyer was held up — by flood-rain — at Winton. He'd tried to phone Daintree to leave a message at the Post Office but the line was out. Not to be deterred so easily, he'd hopped straight into his vehicle Daintree-bound. Obviously, he hadn't got our phone message.

The air was electric. You could almost hear the brainwaves crackle. What about the landslip? There was a 'dozer there. They'd cleared it. The rest of the road? No worries.

The poor unsuspecting fellow had barely time to finish his coffee when my friends loaded me and my luggage into his trusty little green Land Rover and waved us gaily on our way south. On the road to Cairns, some considerable time after we left Mossman, my rescuer confided he'd had a puncture on the way up. We had no spare.

"But don't worry. I'll get it fixed as soon as we get to Cairns."

Oh yeah? We got a second flat. Fortunately not far from a roadhouse and a passing motorist stopped and took Jeff and one wheel back for repairs.

Much later I reached Uncle's place and he, at least, was pleased to see me.

Next day was antenatal clinic day so, mindful that I hadn't yet seen a doctor, though fairly confident I was pregnant, I fronted up.

The doctor was a trifle scathing.

"You've left it a bit late, haven't you?"

I was tempted a week later to use the same words to him when he arrived in the Labour Ward some time after our son had arrived. We'd done very well without his presence so forbearance came easily.

As the nurse wheeled me back to the ward Uncle appeared with excellent timing and his traditional new-baby gift, a bunch of pearly-white rambling roses from his beautiful and much-loved garden. He also had a newspaper cutting which he carefully unfolded and read to me.

A woman in the U.S.A. had given birth to three children all born on the same day at two-yearly intervals. Young Lee arrived on Laura's second birthday.

"You can't let her beat you!"

I declined the challenge to beat the American's record. From now on, someone else could Populate the North. I was bowing out.

3

A Chain is Built

Some bushie once remarked that he didn't know why women want to build a house — it was akin to a dog building its own chain. True. With five small children to educate and care for, I was almost permanently 'tied-up'. The carefree life of the mustering camp was gone. Pressures of responsibility quickly enhanced the aging process.

Laura was allergic to cow's milk. She outgrew the sensitivity after a few years but before she did we had few dull moments. The children were breast-fed until weaned at nine to twelve months depending on how difficult it was to persuade them to drink from cup or bottle. Naturally, with babies, milk is an automatic first choice when menus are planned. Farex and milk. Junket. Egg custards and so on. This greatly compounded the problem until we solved the mystery.

I spent Laura's first year as general manager of the Daintree property, we had just acquired — Woondoo. Not that it took much managing as Lucy and Gerry, two part-Aborigines brought up by the Fischer family were still living there and knew all there was to know about handling coastal paddocks and cattle. During the year, Gerry married the next-door neighbour and literally moved house when she left. It was loaded piecemeal and re-erected downstream at her new home. Lucy went to town to care for Louis Fischer who sold us the property but before they left they broke-in another Daintree stalwart, Billy Burchill, to take over their chores. A more loyal friend would be hard to find.

There was in Daintree township a small school attended by the local children including the 'Cobb kids'. The Cobbs were related to Louis and lived almost opposite on the 'right' side of the river — the one with road access. It was at their place that I left my suitcases when Lee was born. There were six Cobb children including twins, Mary and Caroline. They rode to school, double-banking on reliable ponies which drowsed through the school hours in a special paddock adjacent to the schoolhouse.

To get John and Nancy across the river and down to Cobbs in the morning and back again in the afternoon would occupy half the day so I elected to keep them on correspondence lessons. Young Billy, not yet school-age, had a marvellous time off-siding for Lucy and Gerry and learned to eat mussels, wockles (a kind of yabby) and all sorts of bush tucker with more relish than he applied to meat and three veges.

I once tried to talk him into eating boiled carrot by saying Lucy found it in the scrub but his passion for bush tucker wasn't blinkered. "She should have left it there," was his disgusted reply as he spat out the first mouthful and obstinately refused the rest.

Friday afternoons were the highlight of the week. We took our completed correspondence work to the Post Office to be mailed to Brisbane and on the way home picked up the pillion riders and ports from the school-pony brigade. These were dropped off at Cobb's and as likely as not a couple of the Cobb children would come up with us for the weekend.

Laura was the only one who didn't flourish in that lush tropical valley. Instead of gaining weight she lost it, was restless by day and sleepless at night. Often Billy Burchill would pad in on bare feet at night, lift Laura from her cot and walk her around outside in the cool until she slept. I certainly appreciated the opportunity to catch up on my own sleep and will remain in Billy's debt forever.

There was a clinic held regularly in Daintree by the doctor from nearby Mossman. I took Laura down.

"Is she teething?"

"Yes."

"Well," he said in a superior that-explains-everything tone of voice, busily putting his things away ready to return to Mossman and dismissed the case. A return visit — still teething — brought no more satisfaction.

Rose Cobb, mother of six, exhausted all her advice. Nothing worked. Finally, very early one morning, Jim Cobb and Rose arrived unexpectedly. Jim decided it was time for action. He was taking Laura and me to a child specialist in Gordonvale. Now! Rose would mind the other children while we were away.

Almost as soon as we walked into the doctor's surgery, he had the answer.

"She's allergic to cow's milk. Get a goat. In the meantime give her as much fluid as she'll drink. Sugar, salt, water and fruit juice."

The sugar-salt mix was vile. I held out no hope of getting her to try it but she took to it like a drunk to his favourite booze and lost her starving, refugee-camp look almost immediately.

The next thing was to find a goat.

"You'll get a good milking doe easily around Rockhampton," the doctor helpfully advised. "Won't take much to air-freight it to Cairns."

But, of course, we knew better. Hadn't we seen mobs and mobs of goats all along the roads both to Cairns and to Mt. Molloy?

Rose, who had never been south of Cairns nor further inland than Mareeba, came goat-hunting with me. We began enthusiastically, full of optimism, but as we visited one goat herd after another our spirits plummeted.

"Pity you weren't here last week. I had the loveliest little doe just kidded. Wonderful milker. And so quiet. Sold her to a man in town."

"And I've got three nannies due to kid next month. Any good to you?"

There were endless variations on these themes wherever we went. The joyful anticipation of finding a congenial cure while taking Rose touring soon evaporated. There didn't seem to be half the goats along the roadsides that we'd expected. It all seemed part of a demoniacal conspiracy.

Nevertheless we had seen a huge long-and-wavy-coated buck with enormous upswept horns, a veritable Titan, tethered near a nanny whose kindly smile almost cancelled his sneer at the roadside just south of Mossman. Anything was worth a try and in desperation we drove there.

"Sorry. We're only looking after Billy and Shirley. Their owner's in the T.B. ward in Cairns. He'd never part with them."

Despondently we drove back, thinking how, if we hadn't been so cocky, we could have had that air-freighted, purebred milking doe of irreproachable pedigree in the back of the Land Rover with us — right now.

I brought my radio-transceiver from Crocodile so I could contact the Aerial Ambulance if necessary. Though we now lived close-in, with the river between us and the road we were still reasonably isolated. If it were fine, Ambulance planes sometimes flew north up the Daintree Valley instead of going inland and then north. Charlie Harriman at the Base added me to the early weather stations on the morning sched. That way, I could not only brief him on cloud conditions but could also get my 'traffic' for the day over and done with without waiting around. When the operator at the Base took my weather I still had goats on my mind.

"You wouldn't happen to know where I could get a goat in milk, would you? Over."

The reply was astonishing.

"Sure do. My mate's got one and he's got to get rid of her pronto — or else!'

The next afternoon Jim and I drove to the township in his tray-backed vehicle to meet the truck that brought mail and packages from Mossman and points south. There, in a beautifully-built wooden crate was the sweetest little goat I'd ever seen and a note saying, "Her name is Virginia. Please look after her."

Once Virginia was installed it was only a matter of days before a message came to say we could have Billy and Shirley 'on permanent loan' on condition they were never to be separated.

I thought of that huge, malevolent, yellow-eyed Billy and his quite decent and pleasant companion, thought of the trouble we'd had getting just one small goat and Rose, the kids (human) and I drove down and collected the odd couple.

With Laura improving almost hourly on goat's milk I thought our troubles were over. They'd only begun. With Billy came two long, hefty chains. One to tether him and one to keep him restrained at a safe distance from the unfortunate person fated to shift Billy to fresh pastures.

In our innocence, seeing we were 'out bush', we tried letting him run loose arguing that it was only the constant tethering made him so belligerent and cranky. A mistake! With those terrible horns he bailed-up anyone or anything that ventured out the door and if given the opportunity would charge inside and terrorise everyone indoors as well.

He pillaged the potatoes, onions and pumpkins in their shed, made mincemeat of the vegetable garden and vandalised the fruit trees with those terrible horns, up-ended more-or-less permanent structures and created a swathe of havoc everywhere before the human Billy and I were able to recapture him. Fortunately both nannies got on very well and were so docile and affectionate they almost made up for their paranoid lord and master.

When Butcher Hill resumed normal operations after the Boss's leg healed, Bill went home. Billy Burchill was promoted to Woondoo's general-managership and demoted, the children and I returned to Crocodile.

Billy, Shirley and Virginia came too.

We were to travel, Bill, me, four (at that time) children, Bill's two dogs and our gear in a small short-wheel-based Land Rover with green metal cab and a canvas-canopied back. A rather tight squeeze, especially as I also had two kittens, one for us, one for a friend. The idea was to leave Billy and Shirley behind — Shirley was dry — she wasn't lactating — and take only Virginia.

Bill was due back in about a month to dip the Daintree cattle and without us to clutter-up the Land Rover he could easily bring Billy in Virginia's crate and Shirley, loose, as Virginia was to travel now, back with him.

The chook house, strongly netted with fine wire against marauding carpet snakes looking for free lunches and midnight snacks, seemed the ideal place to keep the two goats not wanted on this voyage. Billy Burchill would release them once we were on our way.

We assembled on the river bank. Rain was falling, the river rising and the tide coming in, backing the water still higher. The two human Billys took one load across in the rowing boat and were returning. Nancy and young Billy Kid stayed with the load on the far side. I had Laura, the kittens were waiting on the sand in an apple carton with specially perforated air-vents and Johnnie was leading Virginia to where the boatmen edged their craft stern-on into the bank when a mini-tornado with two deadly upswept horns and a great deal of righteous indignation hurtled furiously down the steep track we'd just descended.

Virginia didn't stand a chance. It was real whirlwind romance stuff. She was swept off her feet — as too, was Johnnie — butted and chivvied up that slippery ascent at a pace I didn't think meek Virginia could ever achieve.

The two Bills and Johnnie flew after the eloping pair. Virginia's trailing light dog-chain hindered her not at all. Billy whisked her and the end of the chain out of reach in the blink of an eye.

The rain came down at a more determined pace and the river rose another inch or two. Shielding Laura as best I could from the rain I watched anxiously to see if they'd rescued Virginia. I had to have a goat. The tempestuous abduction took but a split second, the re-capture was leisurely in the extreme.

"Mu-um!" An alarmed cry from the other side. "The kittens!"

Just in time, I swooped as they were floating off the sandbank in their apple carton ark on the rising tide.

Eventually the three humans re-appeared with not one but three goats. Billy, with his great horns and unrestrained wrath had all but demolished the chook house. The perpetrator of the hi-jack and preceding vandalism was suitably chained but, with Virginia, the Pearl of his Desire, by his side he came as meekly as Mary's little lamb.

We had expected the Rover back to be crowded with one goat, two cattle dogs, two kittens, three children and the gear but with two extra goats — one of them the notorious Billy — space was extremely limited.

The two dogs jumped in first and perched on the metal 'seat' along the passenger side; the three goats ranged along the corresponding seat on the other side. The opposing parties glowered at each other, all waiting for one of the opposition to make a false move. John and Nancy acted as a peace-keeping force in the middle No Man's Land. Young Billy Kid summed up the situation and decided it was safer on my lap with Laura and the kittens slept on peacefully in their box under my feet.

I was infinitely relieved that outright war did not occur — an odd skirmish was all — but Bill was completely humiliated when we met an ex-Peninsular ringer, a mate of his, at the Molloy turn-off and reluctantly stopped to talk.

Curious about the strange noises coming from the canopied back, Toby put his head inside the window for enlightenment, withdrew it swiftly and laughed uncontrollably for minutes. For years, he brought the shameful incident up whenever he met Bill and relished expanding on the finer details to any audience he could find.

Fortunately, Billy (the goat) made only one more — his final — trip. I doubt if I could have persuaded Bill to repeat his goat-transporting act even in the direst emergency. Shirley and Virginia were good travellers but their presence did mean doubling the usual amount of biscuits taken to be eaten en route. The nannies developed a passion for custard creams and Iced Vovos as well as the humbler homemade Anzacs and jam-drops and thought nothing of snatching them from an unsuspecting small hand when temptingly convenient.

Eventually I discovered a chemist who sold dried soya-bean milk, which did away entirely with goat travelling-companions. Billy and Shirley's owner wrote to say he'd probably never leave hospital and we could buy the pair if we wished. I couldn't get the cheque away to him fast enough. Once legally (and morally) ours, the non-separation clause in the loan agreement was null and void. Billy was up for grabs.

Luckily the reputation of his imposing size and stature superseded that of his evil heart and May Callaghan, with a lifetime's experience with the animals, said she'd love to have him. The speed with which I got Billy to Laura township to meet the Palmerville truck equalled the speed with which I'd sent off the purchase cheque. Light of heart, we sang all the way home.

Virginia died from snakebite after giving birth to twins which we reared on cow's milk and when Laura no longer needed goats' milk to survive, Shirley and the then-adult twins were tearfully farewelled as they left to make a new home with a woman who sincerely loved all goats. Perhaps even one like Billy — though I doubt that would be humanly possible.

When I returned to Crocodile with Laura, John Barry was still with us, doing odd jobs and enveloping us all in his special kind of cheerful optimism. Calamity and disaster could not be sustained with John around. His practicality swiftly found a workable solution and the radiance of his goodwill banished the heaviest clouds of gloom. He was a man with many friends and no enemies.There was nothing about him even the most pernicketty fault-finder could find to dislike.

Returning with mates from Coen Races at the outbreak of the second World War, John and his friends intended to ride to Cooktown and from there go on to Cairns to enlist. John didn't make it. The party spotted a dingo on their ride and took chase. John's horse stumbled and John was thrown to the ground, his neck broken. After a lot of trauma and innovation in that almost pre-motorcar era, he was flown to Cairns.

Doctors held out no hope of him ever walking again and many long frustrating months were wasted before friends suggested he enter a clinic on the Atherton Tableland run by a European genius who specialised in therapeutic massage. Within three years John was walking.

John resumes his old job of ringer three years after breaking his neck chasing a dingo.

From that he progressed to riding and regained his old job of ringer in the stockcamps. It was a matter of pride that he had to take the lead if a beast broke out of the mob, especially if the going were rough or the timber thick as hairs on the proverbial dog's back. As John was at best a little lop-sided, his rash heroics at times led to misfortune. Pride caused quite a few falls and John soon gave up counting the number and diversity of bones broken. 'Broken all the bones in me body' wasn't too far from the truth.

Of course, when there was no work to be done John could party-on with the best of them in Cairns or wherever, with the result that the fate of his innards was similar to that of his bones. A few coils of intestine, pitted with ulcers, were sacrificed to the surgeon's knife. A kidney followed suit and his liver developed pits and potholes but through it all John's cocky spirit and extraordinary cheerfulness never faltered.

One of his friends, Alan, as independent-minded as John himself, was partially crippled by polio. Like John, Alan never considered that the world owed him a living and worked tin mines at the head of the Normanby River, often on his own. A good cook — like John — he often took cooking jobs in the bush and in Cooktown. Both had pleasant baritone voices and sang while they worked. It was a pleasure to hear them. Of course, there was an element of competition which spurred both on and, when extremely piqued one would refer scathingly to the other as 'that hoppy-legged beggar.'

After John had been to town on a bender he'd return sheepishly to his quarters discouraging well-meant offers of assistance. His suffering, he considered, was self-inflicted and should be dealt with by himself, privately. It was hard to dissuade the children from visiting their friend but John's heart-wrenching moans soon sent them scurrying homewards. Ministrations in the form of strong tea or coffee and dry biscuits were rejected politely but with unbending firmness.

On the second or possibly the third day John's return to the world would be heralded by song as he limped over to the laundry with his swag blankets and dirty clothes to wash. As cheerful as ever he was ready to resume normal relations over a pannikin of tea while his clothes gyrated in the washing machine.

"That was a close one!" he'd say. "Old Nick nearly got me this time."

John was determined he'd outwit the devil. "Nick won't get old John," he'd boast. "I'll go when I'm good and ready."

John and Billy (the goat), whom John liked to fancy was one of the devil's creatures sent to torment him, didn't get along. But who did get on with Billy? With John's precarious balance — or lack of it — Billy's rambunctious behaviour was dangerous.

One night in the middle of a particularly bad electrical storm with chain-lightning crackling straight to earth and thunderbolts booming, John decided to visit the outside loo. As he reached the door a grey apparition sprang at him and vanished but not before it knocked him into a crumpled though physically-undamaged heap.

Now fully awake, John got up to see Billy glowering at him, orange eyes ablaze, from the shelter of the shed. He'd snapped his chain and taken refuge from the thunder and lightning in the outside convenience station. It would be problematic who got the bigger fright.

"I reckoned old Nick had me that time," John told us at breakfast.

My mother was visiting at the time and her sympathies were definitely with John. I'd left her home a few days earlier to mind Laura while the other children and I saw to some outside chores. We returned to find my mother, with Laura clutched tightly in her arms, on top of the kitchen table. Billy was in the store-room doorway, a foot or two from the table, blocking off any advance (or retreat) with his menacing horns. The carton of apples Mum had brought with her was practically pulped as Billy munched through them systematically, adding a raw onion now and again for flavour.

"I chased him out with the broom and shut the gate. Twice." My mother complained. "I put a rope around it the second time and he still got in."

She needn't have bothered. Dear Billy had put his horns under the six foot netting fence and lifted up the whole section from the gate to the corner. You could drive a herd of giraffes under it with ease — well, almost, they'd have to duck their heads just a little at the ends.

After John finished the work at Crocodile he moved to Maitland Downs and did up their saddlery gear — a long promised visit — before retiring to the Laura pub for a spell. He had plenty of mates and the days were spent socialising, drinking and card-playing. The nights followed on similarly. On one night, no different from any of the preceding ones to outside observers, John said a cheery 'Goodnight' at his usual time and sauntered off to bed.

Hardly was the lighting-plant motor stopped and the lights out when there was the loud report of gunshot.

John was gone.

The doctor summoned from her home on a distant property was astounded that our friend had been able to survive, let alone enjoy life, with the massive internal deterioration of his body.

John made the final decision. Old Nick was cheated. Our mate went when he was good and ready.

4

Two Girls for all Seasons

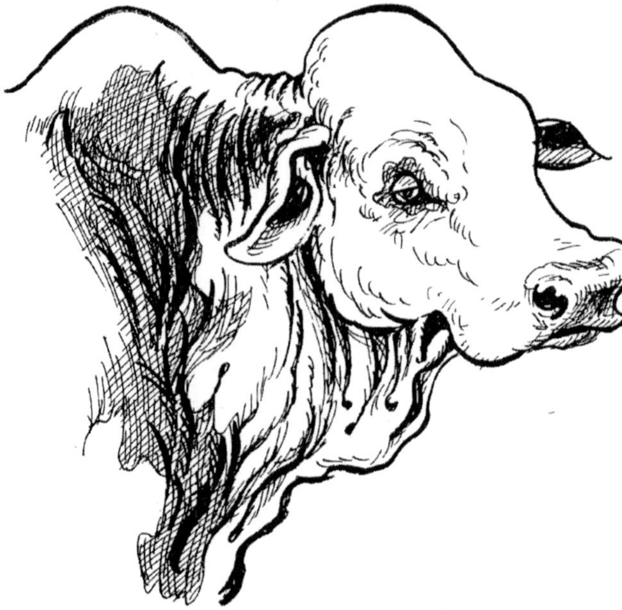

Soon after we returned from Laura Races in 1962, the year Lee was born, my life changed, rather unexpectedly and definitely for the better.

Janice Gowell, then a girl of sixteen or seventeen, away from her home near Brisbane for the first time, was governessing at Maitland Downs. Coming from a large family Janice was experienced with children and knew how to get the best from them -even if she wasn't much older than her pupils. She was finishing at Maitland. Her mother wanted her to return home.

With me having another mother's holiday (a trip to the maternity ward) John and Nancy were back at the Coen Hostel. I decided to drive to Coen, pick them up, and visit Hardy and Iris at Merluna for the August holidays before returning them to Coen to complete the school year.

The Coen Hostel cum cottage hospital was opened in Coen only a couple of years previously by the then Australian Inland Mission — now the Uniting Church. Padre Colin Ford and his wife Margaret 'patrolled' a large area of the Peninsula and were immediately accepted by the scattered community. Colin had the concept of an A.I.M. hostel where children could board under the care of A.I.M. nursing sisters and attend the tiny school at Coen. The sisters could also treat minor injuries in consultation with the doctor at Cairns Base Hospital over their two-way radio, 9CA. Such outposts had proved their worth in other isolated areas and Coen was no exception.

Many mothers, faced with the task of supervising their children's correspondence lessons, had very little formal education themselves. They accepted the new option with alacrity.

With a baby and two rather small children I wasn't sure that it was such a bright idea to go so far — halfway to the tip of the Peninsula — without a travelling companion, though, being racetime, there should be a few local vehicles on the road. If I missed that race traffic, broke down and was unable to get the Rover going again, it might be days before anyone came along. There wasn't that much you could do until someone turned up.

"Would you like to come with me," I asked Janice, "before you go home?"

Janice, eager to see as much of the Peninsula as she could, signed on for the trip. I hadn't seen the two older children since they'd gone to the hostel at the beginning of the year — before my wet trip to hospital — but, each Sunday, the Sisters organised a Home Session on VKA, the Aerial Ambulance radio. Anything newsworthy was usually related by the first children on air but courtesy demanded that everyone had a turn. By the end of the session I'm sure the mothers (and an occasional father) could have recited the news word-perfect.

In the afternoon the children wrote their home-letters, a confirmation of the morning's bulletin. Fishing at the bend of the river just outside the township was a favourite pastime so Sunday's letters were almost always in the following form:

Dear Mum and Dad,

How are you?

We are well.

We went fishing

Down the Bend.

Geoffry Fry (with an occasional variation)

Caught a fish.

No more news,

Your loving son/daughter......

Probably a P.S. to remind parents not to forget to send more clothes, more money or more food and a query or two to see if a special pony had her foal or a pet among the milkers, her calf.

At least it kept families in touch and the children knew to the last small incident what was happening at home. John, our eldest, was only nine and many of the children were younger having been at the hostel from the start of their school years.

Coen Races coincided with the time when parents picked-up their children for the mid-year holidays so, after the races, with Johnnie and Nancy, we set out with Hardy, Iris and their children in a two-vehicle convoy for Merluna.

The road was, like all Peninsular roads, slow. Large stretches of sand could only be traversed in low gear, often four-wheel-drive. As we neared Merluna homestead, Hardy decided to branch off to check on some cattle. Iris and the children were transferred to our vehicle as guides.

Eventually, the iron- roofed homestead came in sight.

"A cup of tea would go well," Iris remarked as we lugged the baby and the gear up the kitchen steps.

Instantly millions of angry blowflies swarmed out to attack and repel the invaders. The kero fridge — as anti-social kerosene refrigerators love to do — had 'gone out'. The small flame that burns to activate the refrigerant gas was no more, extinguished by suffocating smoke and soot which left in their wake blackened walls and ceiling. Before leaving for the races, Hardy killed, taking fresh beef in for the race-camp and leaving behind a fridge stacked full of meat.

With no refrigeration, the meat went definitely 'off', swelled up and forced the door ajar. The blowies needed no formal invitation. Hurriedly we carted load after load of the putrefying meat, maggots crawling up our arms, squadrons of vindictive blowflies in close attendance, to a strategically-placed dump well away from the house and prevailing winds.

All thoughts of cups of tea fled. Our picnic lunch, eaten in happy circumstances on a shady creekbank, threatened to return with every slimey wriggle and foul whiff like Banquo's ghost at Macbeth's table.

We were, of course, fortunate the house hadn't burnt down. Kero fridges are notorious arsonists.

Naturally, Hardy arrived after the fridge was emptied. But the smell persisted and he had the unenviable task of taking the cabinet to pieces, removing all the rancid insulation and leaving it to air in the sun until it was almost free of odour. When it was re-assembled frequent dousings with vinegar and water and a few drops of vanilla helped quell the smell — a little.

Salt beef and eggs figured prominently on the menu until the young ones helped Hardy and Janice get another 'killer'. The malodorous welcome was soon forgotten. There was too much to do and see.

We drove to the old goldmining settlement of Wenlock, its few remaining buildings taken over by white-ant squatters. The old gravestones with sad inscriptions telling of early deaths set us to thinking of how tough it had been in those days. Beside a long billabong, those earlier settlers had the foresight to plant a clump of mango trees with canopies now as capacious as circus tents.

It was a return to the late twentieth century when we reached the new mining township of Weipa. The wharf was a-buzz with activity. All hands and a lot of modern machinery were loading a trial shipment of bauxite for Japan.

On the way back, we made a detour to the old mission site on the Embley to see Percy Wallace's grave. Percy was Sandy Wallace's son, a cousin of the Boss's. He was alone at Merluna when he became seriously ill with fever. With no help available there he decided to ride to the Embley mission. Unfortunately he didn't quite make it. They found him dead, buried him at the mission and his family sent up a stone to be erected over the grave. The mission community there was prone to epidemics of fever and this was one of the reasons it was shifted to the healthier coastal site with its bright terracotta cliffs and cornflower-blue sea.

Percy came from an unlucky family. Another brother, Roy, the Boss's favourite cousin, was killed near Cooktown when a horse stumbled and fell when they were mustering. Both Percy and Roy were in their early twenties. A third brother, Francis, suffered a similar fate to Roy aged only 17. His uncle, another Francis, also died young and his grave at The Range back of Townsville is reportedly the oldest marked grave in North Queensland. One boy, Lex, and a daughter, Marie, survived.

Once the holidays were over, we left the 'hostel' children in Coen. They didn't seem too heart-broken and were soon catching up on all the holiday doings of their schoolmates. Janice and I drove home.

Driving down in relative comfort in the Land Rover, pulling up whenever the notion took us to boil the billy and attend to Laura and Lee, was certainly more congenial than my previous trip behind a mixed mob of Merluna droving cattle. Then we travelled at anything from two to eight miles in a daylight-to-dark day. In the short time Janice had been with me she had made herself absolutely indispensable. I constantly wondered how I'd ever managed without her and as the time came for her to return home I despaired that I'd ever manage alone again. Janice seemed equally reluctant to part from us.

"Do you think," I suggested one night as we sadly contemplated the separation, "if I wrote to your mother she'd let you stay?"

Without much optimism Janice said, "You could try."

Never was a letter so carefully worded. The re-writes were almost endless. But it had the right result. Janice could stay. On one condition. She was to go home for Christmas and then a decision would be made, after due consideration, whether or not she could come back.

She did.

We shared the chores more or less on a fifty-fifty basis but there was no rigid division of labour. Whoever was free at the time attended to the job that, at that particular moment, needed attention. In theory, I did the cooking and washing and taught John and Nancy in the new school year.

The Queensland Correspondence School was a marvellous institution and apart from the educational side of things, they seemed to understand a little of the idiosyncrasies of living in the bush. They didn't make us wait until the new school year began to send the lesson papers needed. If they had, we probably wouldn't have received them until what would have approximated the end of First term.

With the end-of-year parcel — the last corrected paper, a Christmas book 'prize' and a letter from Teacher — came an eight-week supply of papers for the New Year. These could be completed at the Home Supervisor's convenience and mailed in the equivalent to one paper per week batches when the school re-opened.

There was then no Distance Education or School of the Air available in the Peninsula, though the superintendent of the Aerial Ambulance, Tom Briggs, did try to gauge support for a local radio structure. In association with a Principal from one of the Cairns schools he canvassed the area to see what support there'd be for a radio session held daily over VKA. Teachers would be on hand in a voluntary basis to help with any queries or difficulties. By this time, most of the children had moved into Coen so Tom's scheme was never tried.

The week's papers consisted of an instructional part, which mothers could study in private before taking on their teaching role, and the section that had to be returned weekly — the children's answers to the questions. Appended to this latter bit was a page for 'free work'. This could and did mean anything at all and was by far the most popular part of the week's school. Budding artists drew long-horned bullocks, generously-uddered cows, calves, horses, chooks and people. Poets wrote verses. And everyone wrote letters to their teachers who were really one of the extended family. The feeling was mutual as the teachers in Brisbane greatly exceeded the bounds of departmental duty in the care they gave their distant pupils.

Working by correspondence was, in some ways, an excellent training ground. While the kids sadly missed out on competition and social graces, they did learn to read and write, and to seek out information for themselves.

Janice began Billy Kid's schooling and did the outside jobs — with helpers — like riding for the cows and doing the milking. We were to share household chores and also the days when someone was needed to help muster. In practice, the dividing line was rather blurry but it must've worked out reasonably even as neither of us complained.

We always tried to keep a week or two ahead with our correspondence lessons as no one could forecast with accuracy what the coming week would bring. After a Christmas break it wasn't too difficult, especially if we'd had a good Wet and were rain-bound for a few weeks, to sit down and get started on the enticing new papers. Because so many baby calves were born at this time, mustering was at a standstill. Mis-mothering of very young calves, when the mother refused to take the baby after it lost her mother-scent in the dust and clamour of the cattleyards, was something to be avoided. The Daintree 'boys' we usually employed went home for Christmas and didn't return until the heavier rain was over and the calves old enough to be mustered.

John and Nancy were in the same grade. Unofficially, I started Nancy off on John's old papers the year before she was due to commence school. Johnnie thought it discriminatory and unfair for him to sit down all day doing schoolwork while his sister went free. It was very much easier to hold his attention with Nancy doing her interpretation of pothooks beside him.

In her first official year she crept ahead of what was required and when they arrived in Coen, a one-teacher school — there being no-one else in her grade — the teacher provisionally 'put her up' to the next one where she coped well enough to stay. It is certainly much, much easier for a mother who herself hated anything to do with school, to teach two reasonably-willing children in the same grade, than two totally unco-operative terrors in two grades.

Being somewhat isolated had decided benefits as well. The children knew nothing of the Rights of the Child, 9.30 a.m. to 3.30 p.m. school hours or public holidays. Of course, there were Christmas, a vague Easter, the never-to-be-missed Laura races and the bullock muster. School was fitted-in to avoid clashing with these major events.

As there was no School of the Air or Distance Education in our area, we weren't tied to that routine — nor reaped any of its many benefits. As long as we returned the required number of papers, reasonably well done, by the year's end, we were out of trouble and were promoted to the next year.

It wasn't until our sojourn in Daintree that the Cobb kids introduced Nancy, and through her the other children, to the delights of Public Holidays. I found Nancy one morning studying most earnestly the calendar on the school-verandah wall.

"All the red numbers are holidays," she informed me with an implacable look in her eye. "Australia Day, Anzac Day, Labor Day, Queen's Birthday and, "to clinch it all, "Show Day."

We had, within reason, to adjust accordingly. I usually know when compromise is the better course. Bullock muster and Laura Races continued to tie for first place despite the additions to the holiday list.

Usually, unless the latter two events intervened, we did have weekends off. Janice couldn't go anywhere so we compromised by a change in program. No school. As little 'inside' work as possible and as much 'outside' activity as we could fit in.

Crocodile, like most of the Peninsula, was phosphate-deficient and as the 'roads' dried out after the Wet, we routinely took salt and mineral blocks out to the cattle each Saturday. The blocks, a mixture of salt and the minerals that were deficient, weighed about 45 lbs (20 kgs) and were placed in a trough made of a hollow log split longitudinally and adzed smooth inside — so the cattle wouldn't get splinters in their tongues. A rough four-posted shelter was erected with a sloping galvanised iron roof and the trough held at a convenient cow-height on two anchor posts or natural forks, under the cover. The cattle relished the salt and a beep of the Land Rover's horn as we drove up soon had their full attention. They'd trot in from all directions.

We'd been trying to spread Townsville stylo, a legume, through some of the more sparsely-grassed country and found one of the easiest ways was to take a few bales of hay, made after the stylo seeded, with us.

The cattle anticipated this treat and came so close that they'd crowd in with a good bit of greedy pushing and shoving to take pieces of hay from our hands. In return, they scattered the seed in their dung and it came up, well fertilised, all over the run.

The precursor of today's multitude of stylos wasn't a native. It made its way unofficially into the North in fodder and, rapidly adapting, naturalised and spread extensively. Because of its success other leafier and perennial stylos were imported and trialled. Townsville stylo was an annual but though it died off after the Wet, cattle, horses and station chooks lived on the high-protein seed in the Dry season. Unfortunately a disease, anthracnose, almost entirely eradicated the old Townsville strain but by then its more disease-resistant cousins had made themselves at home.

The salt-runs weren't limited merely to taking out the mineral blocks. They were a perfect excuse to leave the house. If the creeks had water in them we swam or fished. Under Janice's patient tutelage the children learnt to swim or, at least, to dog-paddle. The more daring dived from suitable tree-trunks into the pool. It was a great way to cool off.

We explored Split Rock caves near the road and marvelled at the centuries-old aboriginal paintings there. In vain we searched the river banks for signs of where the black artists gathered their ochres. Some of the Main Roads men found ochre patches and gave us the directions required but they eluded us.

Near Split Rock were the caves, blow-holes and arches at the top of the Kennedy hills. These caves were composed of millions of river-worn pebbles cemented together with pastel-coloured sandstone mortar. Pink, blue, creamy and greenish. The children — and Bill — were rather dubious at the theory Janice and I put forward that the hills had once been the riverbed and that water had worn the pebbles so smooth.

Another job we had as relief from 'home duties' was taking fresh supplies to the mustering camp. Bread, cooked meat, cake and biscuits, odds and ends they told us to bring, forty-four gallon (two hundred litre) drums of water, brands, ear-marks, ropes — whatever.

Filling the drum with water was no problem. We put it on the Land Rover empty and filled it with a hose, usually with someone on guard to yell, "It's full! Turn the tap off!" before too much of the other gear was soaked. Getting the full drum off was a different matter. Filled, it weighed in excess of 200kgs. Together Janice and I could manoeuvre it out by 'walking' it from side to side in arcs to the tail-board and giving it a hearty shove, ever mindful of Myles Gostelow's crushed ankle when a drum he was rolling up onto a vehicle rolled back.

Once on the ground a hefty heave-ho in concerted effort and it was standing, bung-up and leak-proof. On the occasions Janice did the trip on her own I couldn't work out how she got the drum off over the lip at the back of the tray on her own and then, to cap the feat, stand it up.

"Easy," she said. "I've usually got the ropes too, so I put a loop of the big head-rope around the drum, tie the free end to a tree, drop the tail-board and drive off."

I was overcome. What radical thought! I'd be scared stiff the almost-sacred head-rope would break.

Strengthened by Janice's success I tried it. It worked. The head-rope was intact and I wasn't struck dead by a bolt of lightning. I'd learnt a new trick, but I wasn't as strong as

Janice and couldn't stand the drum up. All I could do was screw the bung extra tightly before I snigged it off — and hope for the best.

Janice was also quite an athlete and often turned catherine wheels or tied herself in knots to amuse the kids and pass the time. A favourite of hers was to bend over backwards to pick up, with her lips, a threepenny bit (a small coin) from a handkerchief on the kitchen lino. Perhaps not terribly hygenic but most impressive.

When this feat was retailed to an eager group of men at the Laura pub, one said, "I don't believe it. It's impossible."

With righteous indignation we glared at him.

"I can't do it," he explained rather lamely and disappeared after Janice gave a demo.

One day Johnnie and I were detailed to take the camp gear to the Kennedy camp. It used to be a packhorse job, following Quartz Creek up, dropping over the fall and following the Kennedy down to the camp. If, once the head of Quartz Creek was passed, we would have ridden from there over the Byerstown Range to the Palmer River fall, we would have found the old Hell's Gates pass to the Palmer. But — mustering was more important and we left the ghosts of the ambushed miners in peace.

I'd been to the camp many times by horse but never in a vehicle. I had apprehensive memories of the terrain and was a little nervous. The men left home that morning with their riding- and spare-horses and expected us to be there with tucker, swags and so on for them to make camp. We later built a musterers' hut there, complete with rain-water tank (no more drums!) but at the time it was a tent-fly job.

"Where do you turn off?" I asked Bill, not feeling overly confident of getting there with the vehicle. It was more ups and downs than flat going.

Bill gave me a scornful look.

"You turn off where that old red cock-horned cow died."

Ah. Bill had pointed out her bones to me and a small remnant of hide on the top side of the wheeltrack road to Laura. Just on top of one of the numerous gullies before the Kennedy River.

Unexpected visitors called and by the time they'd been refreshed with tea and local gossip, it was rather late. Those men would be looking for their tucker and their swags. Still, all we had to do was to turn off where the cow died and we'd find the road.

"It's plain," Bill assured me. He'd done one trip in and out again with wire for the holding paddock they'd built a week or so earlier.

Johnnie and I managed to ooze a sort of pseudo-confidence as we sped along the familiar dirt road. Suddenly, we came across evidence of Main Road's activity. They'd begun clearing the road, a good chain (twenty metres) on either side of the track. And, as usual, right in the middle of nowhere.

With a decline in spirits we drove slowly over gully after gully, Johnnie acting as spotter, until we reached the Kennedy. We'd gone too far. No sign of the bones of the red cock-horned cow — nor her hide.

Janice picking up a coin with her teeth off a hanky on the floor.

A U-turn, made easily, even with a Land Rover, in the wide clearing, and back we went with me spotting. No luck.

Johnnie climbed into the back in the hope that from the higher vantage point some sign of the deceased red cow would be revealed. It wasn't. We tried a gully-by-gully approach both agreeing it was one of the gullies closest to home. By getting out at each gully and prospecting in the pushed windrows we found a few bones and a cock-horned skull at the third gully. Pay dirt.

We then had to manoeuvre over and around the pushed-up line of timber. As to the 'plain road', there wasn't much choice. Only a limited number of trees were wide enough apart to drive through. We knew we had to go in a general up-stream direction as the camp was just off the river but it was getting late and the sun was forsaking us.

Our spirits soared with hope in a more open stretch and crashed to earth as we came to a sudden drop in the 'road'. 'Gully' would be no true description. 'Chasm' or 'abyss' would be more realistic. The side plunged down at right-angles and managed to come up again barely the length of the vehicle ahead at the same ninety degrees. This couldn't be the good, plain road, surely?

Johnnie was despatched out, with matches this time, to look for a more likely alternative. There was none. Just as I called my fatalism to the fore and put the Rover into four-wheel-drive Johnnie called, "Their campfire! Look!"

Sure enough, there was a fire. It flickered invitingly a couple of hundred yards ahead. They'd know we were at the gully so if it wasn't the correct crossing someone would have come down to re-direct us. Wouldn't they?

We bounded down in a series of bumps and grinds while the water-drum tested the rope attaching it to the frame behind the cab. Up the other side in a series of leaps a mountain goat would have been proud of — and there we were.

"You're a bit late," commented the welcoming committee as they hurriedly unloaded.

"Told you the road was plain," they added. "Must've been. You could do it in the dark."

They'd seen the removal of the red cow's remains but knew I'd remember the turn-off.

Gallantly Bill offered to drive back over the chasm. I thought I detected a crooked half-grin in the flickering firelight and heard a snigger or two from his young offsiders but graciously I accepted his offer. I'm not the adventurous type.

We crossed the ravine, said our goodnights, Bill walked back to camp and I made a mental note to find that wretched cow's cock-horned skull and tie it to a prominent tree at the turn-off.

Good thinking, but Bill or one of the boys beat me to it. It was plainly showing the way to travellers from a convenient tree-fork the next time I passed that way.

Some drovers were still overlanding cattle to the Mareeba saleyards. Peninsular Freighters, a Burton family operation, was poking its way north with semi-trailers and trucking singles out from the more accessible places. The Bamboo Range however caused problems and trucks couldn't climb it without help but semis were shifting cattle from around Cooktown and Laura to the saleyards or the Bacon Factory at Mareeba and to Queerah in Cairns. With Crocodile stocked-up it now took all Bill's time to look after and he gave droving away. The horses and plant were easily made use of on the property.

The mobs of cattle coming down from the Peninsula were getting smaller and more suited to the difficult stock route. Six hundred or so was the norm, usually composed of stores from two or three neighbouring stations, though Rokeby still put off two large mobs of their own steers each year. The Elmes usually took them down and until Southedge was sold to Maurice de Tournouer they were fattened there. Southedge is now the home of Quaid's Dam hated by greenies but admired by mostly everyone else.

Properties were changing hands. With Weipa and the new road, people were discovering that Queensland extended north past Cairns and there were bargains to be had up there. Sales were made at so much a head, from the equivalent of $6 to $10 per beast, and 'the station thrown in'.

The new owners usually mustered, sold most of the cattle to pay for their outlay and put the place on the market again with a nucleus of breeders and plant, and at a price equal to — or greater than — what they'd paid for it, stocked. Civilisation had arrived and with it came the odd millionaire and the American investor. Properties of some two thousand square kilometres and with six thousand head of cattle — plus plant — seemed a steal at the equivalent of $60,000 and there were quite a few takers. Likewise, the vendors were dazzled by their instant rise to wealth.

While the droving mobs came through in the early part of the year it was our job to meet the cattle at the Block Fence — a fence running from range to range at a comparatively narrow place to 'block' stock — and to 'see the drovers through'. Translated, that meant checking the mob to see if any of ours had wandered in and had not been dropped off. Fresh cattle love nothing better than to 'box up' — and sometimes 'box on' -with road- or travelling- cattle. Usually they tire of the game before they leave their home run but at times they continued with the travellers, often unnoticed. As the mob strung through the Block Fence gate, any 'strangers' could be worked back to the tail of the mob where a horseman prevented them from following and the sliprails were shut on them.

It had been raining for days. Creeks were running bankers and the country was boggy. Bill was camped-out and before he left he told us that, at the first break in the weather, two of us were to ride down below the Block Fence and pick up some mares to go to the stallion. These mares were on no account to be allowed to go faster than a walk. Someone was to ride in the lead at all times to regulate the pace.

It was decided Nancy and I should go. Janice would look after things at home. We set off and soon found the mares grazing where Bill said they'd be, on a summer-grass flat at a creek junction. So far so good.

It sounded simple but one old dear — Bill's favourite, naturally — wasn't happy with the pace I set. If she had to leave that luscious creek flat, she'd leave it with a protest. My horse was soon in a sweat trying to keep her on the straight and narrow, then, changing tactics, at every gully, head high and tail flying high over her back, she'd bound up the bank trying to get ahead. It was clear something had to be done — and fast.

We hadn't brought a halter as there were several mares in the mob but we hadn't realised the fresh green grass would make them quite so recalcitrant and frisky. If we had realised earlier, reinforcements enough to make a mobile human yard would have been readily available.

The longest thing I had was my whip but once circled around Bangle's neck it left nothing to lead her with. No time to delay. I took my jeans off. By tying one of the legs around her neck I had enough freeway in the other leg to lead her — and my whip to extend it if necessary.

Once caught, Bangle was quite content to walk sedately at my side with her mates dutifully bringing up the rear. I'd barely had time to congratulate myself on my admirable lateral thinking when all the horses propped, paused momentarily, ears pricked, heads turned towards the Block Fence drovers' camp whence came the unmistakable sound of horse-bells.

We'd heard the two mobs due down from the Peninsula were held up north of Laura and weren't due for days. The belled horses were a mystery. A careful reconnaissance from a safe distance showed it was just a plant returning — horses and perhaps two men. Quietly we turned off into the river, praying the mares wouldn't whinny, silently rode over it at a fordable crossing, through a second set of rails on that side of the river and headed warily for home as noiselessly as we could, like unarmed cowboys in Indian territory.

The sun was shining beautifully as we put the mares into a small holding-paddock ready to go into the larger stallion paddock the next day. I decided to have an early shower and wash my long hair. It'd be an ideal chance to get it dry before dark.

Janice was in the garden cutting the Wet -season overgrowth of weeds with a cane-knife. The children were stacking the cut greenery in what was supposed to be a compost heap. I sat on the steps watching them and sipping tea as my hair dried. In the kitchen behind me someone was moving things on the table. Mentally I checked what I'd left there. Nothing breakable that I could think of.

Then I heard a spluttering cough. Action! I was just in time to see Lee put down the grimy green-pea tin used to store fire-lighting kerosene. Billy Kid had taken it from its high shelf to 'sharpen' his toy pocket knife on an emery stone. I'd forgotten — or not noticed — it.

Lee was holding his breath and looking decidedly blue. I didn't think he'd drunk much, if any, of it. By the sounds I'd heard, he'd gagged on his first sip. Janice said he'd had a drink of milk just before we got home and I had vague memories of milk being the 'antidote' for kerosene.

Almost on cue, the sun clouded over and the black rain clouds rolled in again.

Janice ran in from the yard, sliced her cane-knife into a convenient stump near the back door, quickly got out the St.John's Ambulance book and looked up kerosene. The radio was in the engine room but it didn't take more than a few seconds to get there. It was already hooked-up to a battery. Lee was breathing again but I thought his lips were still cyanosed.

The static was bad.Time after time I called VKA getting no answer and was just about to give up and try something else when a voice came on — the one I didn't want to hear. Its owner was a morse-code expert and knew nothing about the bush.

"Receiving you faintly, Crocodile. I can't read you." It was very noisy with atmospheric interference. "I think you have a medical. Try again in the morning."

I unhooked the battery without bothering to reply and went back to Janice. From the St.John, she'd found the milk was a lucky break; that we shouldn't try to induce vomiting and that pneumonia could follow. That, unhappily, we already knew.

We had one stallion, just returned from being loaned to a friend, locked in the cattle-yards relying on water from a hose, the mares in the holding square with no water — they'd drunk at the river — and the other stallion in the paddock adjoining the mares. He was already showing a keen interest in them to the chagrin of the fellow in the yards. Added to that, just about every article of clothing we all possessed was on the line and not quite dry. Janice had taken advantage of the sun, too, to wash.

With storm clouds gathering there was no time to waste. Grabbing a few things we piled into the Land Rover heading for Butcher Hill. They'd recently installed a telephone — one of the first in the area. I'd ring Matron Brown in Cooktown. She'd know what to do.

By the time we reached our house-paddock gate, Lee regurgitated a little kero-perfumed milk, regained his normal colour and was enjoying himself no end. Nothing like a trip to see Grandad. I had second thoughts about going further but cautiously decided I'd phone Matron to be on the safe side. She'd tell me what danger signs to watch for.

Though she could wither you to a dessicated chip with a sidelong glance and turn your legs to jelly with a few terse words, she knew all there was to know about emergencies and practical nursing. Her medical knowledge was backed by a lifetime's experience and commonsense. She'd never failed anyone that I'd heard of and was always calm and in control.

We were out of luck. Matron was away.

"Bring Baby in," said the Sister in charge over the phone line.

I was hesitant.

"I'll ring the Doctor in Cairns for you."

Doctor duly came on the line and repeated the Sister's request. Baby would have to have his chest X-rayed.

Our Daintree nurse friend, Irene, was at Butcher Hill with Wadgee. Though she thought Lee looked rather healthy she advised he should have his chest examined. She'd come with me.

There was still some daylight and with two drivers we might even be able to come straight out after they'd checked him out. That decided it. We left. Janice and the other children were to return early next day in a borrowed vehicle to hold the fort at home, sort out the stallions and mares and generally keep the peace. We'd be back as soon as possible. Might even beat them home.

All went well until we got to Barrett's Lagoon, a noted home of salt-water crocs though in those days shooters kept them somewhat in hand and they tended to be more timid. The road became progressively wetter and more slippery and the heavy clouds completely blacked-out the sky.

The headlights picked up a vast expanse of shining water. The lagoon was across a low section of the road and loomed ahead of us for a considerable distance like a forbidding inland sea. With more caution than daring, I stepped out to test the waters. The slightest movement and I reckoned I'd be back in that Rover before a croc had time to line me up.

"I'll shine the torch," offered Irene.

We had the headlights but the torch was more discriminating. It could be shone on anything suspicious out of the range of the headlights. I don't know why I didn't carry it though, as when I edged nervously forward, Irene was shining the torch everywhere but on my path ahead.

Neither did it help to hear her gasp, "Oh! What was that?" every so often and flick the torch upstream where something moved. From the ripples and splashes there appeared to be a lot of aquatic life out to greet us.

The water wasn't deep — just over my knees at its highest — so I turned at a pair of guideposts and returned — at a far greater speed than I'd ventured out. From the posts the road should have risen gradually as they marked a shallow culvert. With the distributor covered Irene drove steadily through in precautionary four-wheel-drive, the bumper and bull-bar parting the water nicely like the prow of a sailing ship.

Overcome by the earlier excitement Lee slept soundly on the seat. Children's car seats hadn't yet made their way into the Peninsula and I'd been holding him up against my shoulder in case he had inhaled any kerosene fumes, though, with the cooler weather we hoped the vaporisation would have been minimal.

Irene and I held our breath as the Rover slowly edged through the water, dipping a little here with a bit of a splutter, re-emerging on a higher patch only to slide down again to rise at the end of it like a clumsy old hippo, from the crocs' den to relatively dry land. Thank goodness.

From then on, the short drive to the Hospital was a breeze.

Two Sisters, one off-duty, met us at the door and took 'Baby'.

"A man rang to see if you were here," one of the Sisters said.

We were mystified. Bill knew nothing of our episode. Sister explained that Sonny MacDowell had called at Crocodile on his way to Laura — it was a good place for a cuppa at any hour. He'd seen the cane-knife in the stump, the opened First Aid handbook and thought the worst. As soon as he reached Laura he rang the hospital to see what had happened and if he could be of help. When Janice returned, all the doors and louvres we had forgotten to close in our hasty departure, were shut and our laundry — bras and knickers included — was heaped nice and dry on the kitchen table.

That's what friends are for.

Lee woke when we arrived at the hospital and didn't object at all when Sister checked him over and put a stethoscope to his chest. He looked fine to her and his chest sounded clear — so far. Smiles all round. We cheered considerably and thought we could re-fuel the Rover and take Lee home.

"But," said Sister, "he can't be discharged without an X-ray."

He hadn't, as yet, been admitted. That was a mere technicality Sister soon remedied.

Though Matron could take the X-ray it would still have to be seen by the Doctor. We'd picked the wrong day for our jaunt. Doctor was in Cooktown the day before for a clinic and wouldn't be back for a week. It would take about the same time to mail the X-rays to him in Cairns.

Did we have somewhere to stay the night? There were spare beds we could use. They'd put Lee to bed straightaway and keep an eye on him.

It was a bit of an anti-climax but it underlines the problem of small children in remote areas. You must always err on the safe side. One can't afford to take risks. There are too many sad stories told and too many mothers haunted for life by those two little words, "If only..."

We knocked up a friend just as she was about to retire and kept her talking for hours though our eyelids kept sliding down like a lizard's and we badly needed the proverbial matchsticks to prop them up. It was after midnight before we went to bed in borrowed night-attire.

Next day, after presenting ourselves early at the General Store where owner Kath Savage lived 'over the shop' and attired in colourful new cotton housefrocks bought-in by Kath by the dozens mainly for the Aborigine trade, we returned to the hospital.

Matron was on duty but our joy on seeing her was short-lived. Lee definitely had to stay until Doctor saw him. She took us to where he was holding court in a small room that served as Children's Ward. He shared the cot with a pile of soft toys and the ward with a boy, a year or so older, from the Mission. Sister was reading to them both. They'd breakfasted, had a bowl of icecream each and Lee didn't seem particularly pleased to see us.

He'd slept the night through, Matron said. She didn't think he'd ingested much kerosene and the bit he did take was fortunately blanketed by the milk. However, Doctor had to see him before she could discharge him. There was no need for us to hang around.

We were on good terms with all the Main Roads and Council men camped near us but one in particular, Leffy Buhmann, was a long-standing family-friend both of our family and Matron's. Her suggestion was to send Lee home with Jeff. He usually left town the day after the Doctor's visit.

"No trouble at all. He'll be right with me," Leffy assured us. "Matron can let me know when she's ready."

Leffy, too, was packed-up and ready to leave for his current week's trip. We said our farewells. Lee appeared rather impatient for us to be gone so he could devote his attention to his new friends. We took the hint and excused ourselves.

Though the water at Barrett's was about the same depth, crossing it in broad daylight was a lot less scary. Leffy told us rather gleefully that Henry Hanush shot a couple of big salties there, spot-lighting, the night we came in but we were safely across and had no intention of going that way again — especially after dark — until at least after the Wet. Thankful for her support, if not the vagaries of her torch-bearing behaviour, I left Irene at Butcher Hill. Back at the ranch, Janice had everything under control and the welcome mat out.

Lee duly arrived home in good spirits and health and a hospital nightie, with Leffy the following week. We never found what happened to his own clothes. Nor particularly cared.

5

Hooked on the Grand Parade

Janice was still off-siding for me when together we decided to try our hands at showing cattle in the led-in section at the Cairns Show.

For some years both Bill and his father had sent cattle down for what was then the 'fat' cattle classes. Cattle were held-over from the bullock muster in May, a single huge bullock for the 'heavyweight', judged on weight alone and others in groups of three selected because of suitability for a special market. 'Export' bullocks and cows were the big, aged cattle required for the British and the U.S. markets. After supplying meat to Britain during the war the U.K. gave Australian exporters a Fifteen Year Meat Agreement and were perhaps our best customers until their entry into the European Common Market. The U.S. market opened a little later and was a godsend. It created an almost miraculous outlet for 'manufacturing meat' for hamburgers. Most popular were the old unwanted bulls that had previously been unsaleable and had no commercial value at all. Bull beef, being dry, readily absorbed liquids and made the very best hamburger patties.

'Chillers' were younger and lighter steers. 'Yearlings' were just that and 'local trade' steers and heifers were more youthful than the chillers and weighed less than the export ones.

Bill usually took these cattle with a few butcher's bullocks to make up the mob, down to Daintree via the China Camp stock route that, being parallel to the coast was a lot shorter than the long way round through Mareeba. It wasn't an easy trip. The number of cattle that could be taken was limited by the mountainous terrain, by the river crossings — the Daintree itself was crossed five times — and the fact that the route went through the 'scrub' as the rain forest was then known. Each year that portion of the route was kept open by someone from the Douglas Shire who cut any re-growth from the tunnel-like path, barely wide enough in places for stock to travel in single file. Stories of pack-horses toppling down a mountainside in their effort to secure a particularly enticing clump of grass were commonplace.

Once in Daintree the cattle were normally spelled for a few weeks at Ned Cobb's paddock until showtime. From there it was a relatively simple matter to truck them down to Cairns.

Now we had Louis Fischer's Daintree property on the opposite side of the river to Ned's we had an ideal base from which to select show cattle. The rivalry was strong between the Boss and us. He won the first Carcase Competition inaugurated by Amagraze at Queerah Meatworks and we won the next six.

For some years also we'd read the show schedule avidly, glancing briefly through the Cookery, Horticulture, Photography and Fancywork pages to linger over the Stud Cattle classes, the 'led-ins'. Apart from the transport possibilities the new road opened up, what stimulated our current very pressing interest was a peculiar section in it for 'cross' cattle. It had nothing to do with mood, temperament or unpleasant disposition. It was an opportunity the cattle committee considered conducive to encouraging would-be Brahman breeders to exhibit what they had — the crossbreds not yet graded-up to purebred status. There was a similar section for Santa Gertudis 'cross'. Brahman and Brahman-infused cattle were beginning to stir interest in the tropics and this was thought to be a way to foster that interest, at least until the number of purebreds built up.

Zebu, Bos Indicus, or cattle with humps and other peculiarities had been in the North a long time. Some had gone to Botany Bay and other outposts in the early years too but their bloodlines were for the most part lost. They were usually taken on in India or Ceylon (Sri Lanka) to provide milk en route for invalids and infants after 'British' dairy cows brought for the purpose succumbed to the harsh conditions. Lord Hopetoun, our first Governor General, recovering from a long illness, came to Australia with a Zebu cow to provide the vice-regal milk. The Zebus made a more lasting impression in their true home, the tropics.

In the early 1900s a far-sighted Curator of the Melbourne Zoological Gardens sent Zebu bull calves, surplus to requirements, to cattlemen friends in the North. These were descended from animals that came direct from India. Our Daintree herd derived from the best of these cattle from the Zoo — the Robbins' bull.

The hybrids handled the ticks and the steamy heat so much better than the more usual Hereford or Shorthorn with their temperate zone background that in 1933 a group of cattlemen joined with a research organisation (CSIRO) to import more Bos Indicus, this time from the U.S. The importation also included a Santa Gertrudis bull.

While various herds bred-up to Zebu and zebu-cross two breeders in particular concentrated on 'stud' zebus. They were Ken Atkinson of 'Wairuna' and Maurice De Tournouer of 'Wetherby'. Louis Fischer didn't try to upgrade to 'purebred'. He preferred to sell cattle to the meat trade and thought a crossbred beast of about three-eighths to three-quarters Zebu had the best carcase.

By the mid-1940s enough interest was stimulated to form an association of Zebu and Zebu-cross breeders which would record the herds and encourage their upgrading. This happened after a meeting in an agent's office in Mareeba and in later years the Australian Brahman Breeders Association, the breed societies for the Australian-bred Droughtmasters, Brafords and Brangus cattle were formed from this original registry. Later imports were made both of King Ranch Santa Gertrudis and beefy American Brahmans.

To join the Brahman Breeders Association in the early sixties a would-be member had to have a token number of registered cattle and to be nominated by an Association member, often the person who sold him (or her) the registered cattle.

If registered cattle (and the where-withal to purchase them) were in short supply the herd could be upgraded by means of a system using 'appendix' females mated to registered bulls. A 'foundation' cow, be it a Zebu with a big hump, long ears but no registration papers or a hairy Shorthorn, was entered in an Appendix A registry. Her female calf by a registered bull, a theoretical half-bred, was eligible for Appendix B. The next female cross, a three-quarter-bred, an Appendix C. The result of the next cross, both male and female animals containing seven-eighths Brahman blood were 'Ds' and to all purposes at that time, 'purebred' and from then on produced 'purebred' calves. As numbers of registered cattle grew the 'pure' status was extended another generation to a fourth cross until finally there was no need for appendices and crossbreds at all.

In the early days it could become mathematically complicated. Both Mr. De Tournouer and Mr. Atkinson had also each purchased Santa Gertrudis bulls in later imports of Brahmans and Santas from the U.S. The original Zebus were almost pure Indian blood and were distinguished as 'full bloods' from their U.S. cousins, the graded-up American Brahmans. As the Santas had nominally three-eighths Brahman 'blood' the fractions became confusing when allowances were made for the extra three-eighths — or the missing five-eighths.

Mr. De Tournouer was meticulously accurate and it was by no means uncommon to see one of his animals described in brackets as a sixty-three/sixty -fourths or even on rarer occasions as fine as so many ninety-sixths or one hundred and twenty-eighths.

Johnnie, who wasn't the best maths student in the North found the fractions a little bewildering until a friend, Bill Edmonds, who was adding a red Brahman herd to his

Droughtmaster stud took Johnnie to his machinery shed and explained it with the help of a set of pre-metric spanners.

"See, Johnnie, half goes to three-quarters," he indicated the sizes in inches engraved at either end of the spanner. Picking up the next size from his set he clinched it with,"And seven-eighths goes to fifteen-sixteenths."

Understanding dawned. They didn't get to the finer points of the ninety-sixths and one hundred and twenty-eighths but from then on Johnnie had a better idea of the mathematical mysteries of upgrading.

Both Messrs Atkinson and de Tournouer were long on pedigrees and one 'full blood' bull we bought from Wetherby had a pedigree long enough to reach Brazil and an ancestor called Tipus Ideale.We imagined the bull in question was what Senor Tipu considered was the ideal bull, and may have been partly right as some learned person later said Tipus Ideale meant Ideal Type. However, many years later when interested in Old Fashioned Roses, I found in a rose catalogue a Tipus Ideale. It was a rose like the florabunda Masquerade with flowers that changed colour as they aged so that one bush could have flowers of several colours.

Zebus in those days came in all hues. There was even a rumour, most likely started by Bos Taurus enthusiasts, that the new Association was offering a thousand pound ($2000) reward for the first green Zebu. That was the only colour they didn't have.

Knowing how Zebu or Brahman calves born red could change colour on occasion to either black or white I wondered if there were a connection between the rose and Senor Tipu's bull's colour scheme.

The first event we attended after Janice joined the family, and before we did the trip to Coen, was the Cairns Show. Normally we travelled with swags and tuckerbox, so that if we wanted to pull up or felt sleepy, we just stopped off on the side of the road, unrolled our swags, boiled the billy if we felt like it, and that was that. It was also the custom to camp with tuckerbox and swag at the sale yards when we had cattle for sale and couldn't do a 'there and back' trip in the one day.

Bill's father's cousin, Eddie Earl, chief cattle steward at Cairns Show was horrified when we turned down his generous offer to stay at his place, 'Balaclava' at Earlville. We couldn't camp out with a young baby. He reached a happy compromise by bringing his caravan in to the showgrounds for our use. A wonderful idea. It was in a quiet corner close to the cattle so we could watch them and, as they'd been re-newing the old wooden cattle-pens, there was plenty of wood in the form of rejected rails for our campfire.

Not long after we arrived, Tom Booth from Glenray, Daintree, pulled up at the ramp in his faithful red truck — of almost the same vintage as its owner.

We should have noticed Tom anticipated trouble when he took the truck out from the ramp after he unloaded and parked it to make a strategic wing leading from the gate at the corner of the fat cattle pens. At his request we lined the Land Rover up nose to tail to the red truck to make an even more formidable barrier stretching almost to the cattle stalls.

By some slight oversight or fit of over-confidence, no one thought to shut the big galvanised-iron double entry gates that led to the street and were on the showgrounds road that separated the horse stalls from the cattle section.

Dust rose from the corner pen for a few minutes, then Tom and his son, a split-second behind him, rocketed out on the end of a pair of lead ropes, jet-propelled by escaping Droughtmaster-power eager to get back to Daintree without delay.

"Block 'em!" instructed Tom and we flew into position in the missing link between the vehicles and the stall.

No use. The bulls, not taking kindly to their first fleeting glimpse of city life were heading for home. Tom lost his bull. It was careering in a wild circle kicking out at the trailing lead rope that tangled round its hoofs. His mate's progress was somewhat impeded by its 'leader' being towed, leaning back like a surf-skier with the brakes full on. We watched with interest wondering who'd win in the battle against gravity when Tom yelled, "Gate!"

With a tackle that would've earned her a green and gold guernsey in the Australian football team, Janice threw herself between the escaping bull and the open gate. He'd taken on the weakest link, young Billy Kid, who naturally stepped aside to let him pass.

Willing hands speedily clanked the huge gates shut while the bull, seeing he was thwarted, stood there eyeing us off in a none too friendly fashion, his sleek red coat quivering with the thrill of it all. Eventually, by means of pulling at one end while pushing at the other, the bulls were ensconced in their stalls for the night.

At Laura Races we'd had a visit from an old school friend of mine, Erica, who'd spent some time earlier with Ruth and me at Butcher Hill. She was now, like us, married and was accompanied by her husband, well known to all bushmen, R.M.Williams. They promised they'd call and spend the night with us at Cairns and timed their arrival precisely for the moment we sat down around the campfire to drink the tea made to settle Tom's nerves.

We introduced Erica to Tom and Barney and Erica said, "This is my husband, Reg." Hands were shaken, tea drunk and Tom got wound up on his pet topic of genetics and how to breed cattle within a 'closed' herd.

We'd heard and read it all so many times our eyes were beginning to glaze over but R.M. and Erica, hearing it for the first time and having read Tom's lengthy pieces in Queensland Country Life, naively encouraged him by asking questions.

Tom's theories involved such unaccountable things as 'marbles in a bucket' (the gene pool), 'seagulls on the beach' and 'fruit flies' (examples of homogenous likenesses within a breed). I'd read some books on genetics where these terms were used but Bill hadn't and he thought simplistically that breeding bulls depended on bulls and cows rather than marbles, seagulls and fruitflies.

R.M. argued that a commercial breeder was the best person to know what traits should be selected to produce the ideal bull. This was heresy to Tom's school of thought and he said so.

"Two entirely different things," he argued and thought for a second or two to find an analogy we would be able to understand. "You can't do two different things with a high degree of excellence. You can't knock off work carrying bricks to- to- to design Dior ball gowns! You'd have to be a genius. And they're short on the ground."

"Why not?" asked R.M. innocently. "I knocked off carrying bricks to make shirts. Do well out of them, too. Sell them by the thousands."

Tom, usually so quick in repartee, half-closed his eyes to look more closely at R.M. After all, the face should have been familiar enough.

"What did you say your name was?" he quizzed, eyes squinted to aid recollection.

"R.M.Williams," we chorused.

Immediately Tom was up on his feet with hand proffered for another 'shake', "You are a genius," he said and we were spared the genetics, marbles and fruitflies while he talked saddles, quartpots and boots with R.M.

My first contact with R.M. came about when I first went to work at Butcher Hill. Besides supplying boots, shirts and saddles to ringers Australia-wide, R.M. also gave them Hoofs and Horns, a monthly magazine, for spare-time reading. A group of ringers met, complained there was too much 'hoofs' (show riding and eventing) and not enough 'horns', meaning the goings-on in the bush. I was deputed to write to inform the editor of our feelings. This I did, signing it 'Six Disgruntled Ringers'.

In the next mail, only a month after sending it, came the reply. The reason there was so little news from the bush was that no one sent any in. Would I like to supply a monthly column? I would, and did for many years. When my antiquated typewriter refused resuscitation and departed its functional life I missed a couple of issues. In the mail came a beautiful portable typewriter, paper, envelopes, even stamps and a note from R.M. saying,' No excuse now. Keep writing.' He was an extremely generous man and I am by no means the only one to have benefited from his openhandedness.

Back at the show, with Lee only a baby, Laura just two years older and Billy Kid at an age that needed watching all the time, I was hamstrung but both Janice and Johnnie found themselves jobs as leaders.

Mr. Atkinson gave Janice his bull, Wairuna Brutus, to lead in the Grand Parade after the judging and Johnnie, who was home for the Show, pleaded to be able to take the younger and quieter of Tom's two in the parade also.

I was a bit doubtful, but they'd behaved well in the smaller arena down near the stalls where they were judged. I didn't then know the hair-raising, spine-tingling ways of Cairns Grand Parades and both Tom and Johnnie were very persuasive.

"Wonderful experience for the lad," said Tom who had urgent business elsewhere and was cognizant of the fact that every ribbon-winner had to parade in the large main ring.

"Can I, Mum? Please."

I gave in.

We all accompanied the cattle and their leaders to the staging point at the ring entrance. It was traditional for the beef cattle to head the parade with the Supreme Champion Bull of the show leading. At his heels, at a slight but respectful distance, came the rest of that winning breed, followed in turn, by the champions and also-rans of the other breeds. Then came the dairy cattle who usually came down from the Tableland for judging the evening before and left directly after the Parade so they wouldn't miss another milking.

Led horses succeeded them with the Supreme Champion Stallion proudly heading the equine cavalcade that ended with the tiniest of ponies and their tiniest of riders.

As a Wairuna brahman won the coveted supreme ribbon Janice was one of the first into the ring with her charge. Johnnie and Barney slipped in quietly among the other Droughtmasters. A couple of ponderous Poll Herefords bought down South by Percy Edwards and exhibited before going out to Abingdon Downs shambled out to bring up the tail.

The dairy cattle, led by their champion bull, tricolour ribbon encircling his more than ample paunch, followed them, spick and span, stepping out smartly besides the white-coated leaders, very businesslike and well-mannered but eager to get back to routine and home in time for that evening milking.

From a box high up at one side of the ring a pleasant male voice gave a commentary on the breeds and the prize-winners as they made their way slowly in a royal progress circling the ring to applause from their loyal subjects, the crowd.

Suddenly the circle was no longer unbroken. The Droughtmasters shot out in all directions like a school of tiddlers attacked by a hungry predator.

My heart thumped wildly as I saw Johnnie's little bull careering as far as he could from the spot where he'd been but a second or two before. Whatever it was that upset him, he wasn't waiting to check it out. There was a good deal of jerking and head-pulling until the leading Droughtmasters were brought back to some sort of control. Not so the two Glenrays. They were going home. It was obvious, in their opinion, they should never have left there in the first place.

Watching, helpless and horror-stricken, I was heartened mightily to see that bull-leading was one skill Johnnie was learning fast: He readily adopted Barney's water-skier's stance and was leaning back on the lead as far as he could, his Williams' boot heels cutting neat little twin furrows in the short green turf.

He seemed to be in no immediate danger so I slipped from mother mode to professional and tried to beam out the message, "Stick to him!" for the most abject disgrace a handler can suffer, is for his or her charge to get loose in the judging ring or during the Grand Parade. Heels digging deeper, leaning back at about a thirty degree angle, Johnnie stuck to that rope as tightly as a barnacle to a ship's bottom. Gradually, the micky slowed enough for some of the other Droughtmasters to join him. With their comforting presence both he and Barney's bull who'd fled to the big, brawny Herefords for protection, were coaxed back to the parade.

Johnnie adopted Barney's water-skiing stance and his Williams' boot heels cut twin furrows in the turf.

They were somewhat on edge, ready to take off at the slightest alert, on the long walk down through the crowd to the cattle stalls, but they all made it.

"It's that bloody loudspeaker," one of the handlers complained. "It waits until you get up close and blares out. Frightens hell out of the cattle."

I had noticed that the circle swung slightly inwards away from that one particular amplifier so thought he may have found the reason. One the other hand the little bull wasn't too happy with city life and mightn't have needed a 'reason' to stage a demo and lodge his protest.

Tom missed it all but Johnnie was above himself with pride. He'd stuck to his bull and come out unharmed with a good story to tell his mates.

It took a while for me to get over the initial shock and to obliterate the vision of our first-born all bloody and mangled by a mad bull's rampage, but recounted around the campfire that night it seemed no-account and positively tame compared with stories told nonchalantly by the old hands. The horror faded and the humour of the situation presented itself more clearly. The look on Johnnie's face as he shot out on the end of that halter was comical, so the others said, and Johnnie was happy to agree. He'd been blooded. He was 'in'.

While my desire to field a show team might have diminished somewhat, Janice and Johnnie's enthusiasm increased a thousand-fold and they had almost twelve months to win me to the cause. As the months went by the memory of Johnnie being snug by the boot-heels dimmed. Janice's enthusiasm was catching and hours were spent studying the show schedules once again.

We were still a year or so off purebred calves but we did have crossbreds. With the crossbred section there in the catalogue, ready, waiting and most inviting, the question was — would Bill be in favour of our scheme?

He operates on the principle that everyone — his family included — is entitled to attempt to do his or her own thing as long as he, Bill, is neither bothered nor sent bankrupt by the experience.

We were more than happy to adhere to those conditions and had two potential show animals in mind. Grub, a first-cross bull calf by one of the 'bought' full bloods out of a red polled (hornless) cow we used as a milker and a Daintree-bred Santa Gertrudis 'cross' bull just under two years of age.

Grub was the simplest proposition. Tied-up and handled in his role of milker's calf, he was well on his way towards leading with a halter, a requisite for all stud and 'cross' entries. The other hopeful, with a superbly docile, yet inquisitive, temperament was completely unhandled except for his sole rather daunting experience with humans at the time he was branded.

Being a milker's calf, Grub was mine. It was traditional that the women, who usually attended to the milking, owned the milkers — which were, at that period, rarely of a recognised milker breed. Quiet 'bush' cows that appeared to have good milk supplies were co-opted for the job. Funds from any sales from the milker herd could be appropriated, with perfectly clear conscience, to buy things for the home and other items not categorised as necessities. The Daintree bull belonged to Bill, or more correctly to W.H. & L.C. Wallace — us.

"If you can get him to lead in time for the Show you can have him," was Bill's magnaminous offer, possibly made with a modicum of doubt that we wouldn't succeed and he'd be spared.

It so happened that Janice had fashioned a leather halter to fit her prospective pupil — just in case. We slipped the halter onto the cooperative, unresisting bull. He was greatly interested in the whole proceeding, smelt the halter, licked Janice's arm and looked perplexed. Humans got up to some funny antics. While the children watched from outside the rails — probably expecting high-jinks and a bit of lively entertainment — we both tugged tentatively at the lead rope.

Miraculously, inspired by an innate team spirit no doubt, half a tonne of bovine muscle sauntered towards us. Braced for stiff opposition and alerted to take the water-skiers' position at any resistance, we nearly fell over backwards as the rope slackened. It wasn't what they'd anticipated but our audience laughed appreciatively.

"Oh, he's wonderful! Isn't he wonderful?" Janice enthused in joyful relief. To be truthful, we weren't all that confident in teaching a two-year-old bull to lead 'cold-turkey'.

He had a name. He'd earnt it. Woondoo (after Daintree) Wonderboy. Grub's transition from milker's calf to show bull wasn't as spectacular. He engaged in passive resistance rather than active aggression. He was due for a name-change too and when he was leading obediently, standing properly 'with a leg at each corner' and wearing a corn-sack rug at night to keep his winter coat at bay he became Crocodile Dorothy's Pride.

His mother, Dorothy, was very proud of him. He was long, thick-set and heavy and the dash of Zebu gave him an attractive sleek coat with the slightest hump and droopy ears. The problem was that Dorothy was so taken with her son's increasing good looks she had to be placed in another yard to stop her licking his hair up the wrong way after he'd been groomed. A short sleek coat was a must.

Our next problem was what to feed our show poddies. Mr. Atkinson to the rescue. Get some laying mash and pollard and give them a dipper-tin each of that, a slice of lucerne hay and all the Townsville lucerne hay they'll eat.

Progress had also come to us in the form of a telephone. In fact, we were an exchange. Our number, Crocodile 1. A friend of ours in Brisbane laughingly reported his secretary's reaction when he asked her to get him '1 Crocodile, please'. The phone became a lifeline for practical advice.

When the powers-that-were decided our road to Laura township was 'all weather', the old Cooktown to Laura railway laid down in the Palmer goldfield days was summarily torn up. Fortunately it wasn't all bad news even though a couple of heavy Wet seasons isolated us all when the 'all weather' road went 'out' fulfilling the locals' pessimistic predictions. The steel rails came back as telephone poles and a phone line was erected from Mareeba to Laura connecting with the old one up the Peninsula and with a 'U' branching off to link-up Crocodile. It was certainly appreciated, not only by us but by travellers as well.

By courtesy of that phone Mr. Atkinson encouraged us through early disappointments to later successes. Years before, he had exhibited, no doubt to an unappreciative audience, the first Zebus at the Brisbane Exhibition. At that time they were about as popular with conservative cattlemen as feral cats are now with bird lovers. The sleek, hump-backed cattle were referred to contemptuously as 'yaks' — great shaggy, long-haired cattle that look like titanic tea-cosies and to which the Zebus bore very little resemblance. Despite this inauspicious debut, Mr. Atkinson still took his select show team to Cairns and often Townsville and the animals' popularity was on the increase.

As the time progressed closer to showtime, Janice assured me and I, in turn, reassured her, that our charges wouldn't disgrace us. That was all we hoped. That they would be up to 'cross' standards — and that they wouldn't pull away. The thought of that happening made us extremely apprehensive. It was small comfort to know that it occurred quite regularly in Cairns even if Johnnie had won his bout with ill-fortune and Tom's bull.

Bill kindly took our two show bulls down to Cairns for us, Janice, the kids and I trailing him in the Land Rover. Our caravan wasn't due until the next day so we settled the bulls in a stall each, making sure the rails at the open fronts were securely tied and made ourselves at home for the night in a lock-up feed-room next to them. It had been Tom's camp and he wasn't coming this year. Maybe the nervous stress was too punishing. Bill, his good deed done, washed his hands of us and went to pick up the real cattle, the 'fats' from Daintree.

Someone had pinched the light bulb from the feed-room but we managed quite well with our torches. During the night, trucks came in with other cattle, unloaded and left, their grooms obviously staying somewhere off-grounds.

A quick survey in the morning showed Droughtmasters, a team of polled Devons from Collinsville, Mr. Atkinson's big Fullbloods, two young bulls from Mt. Garnet and Alf, "Pop", Westman's team from Central Queensland. No 'cross' cattle. We didn't know whether to be pleased or sorry. Grub and Wonderboy would have no competition.

Pop Westman had a caravan on the grounds. It came out next day with ours and soon we set up a congenial relationship sharing cups of tea, borrowing forgotten items and, as far as the children were concerned, gorging on the huge watermelons provided by Pop.

Pop didn't like horns. They were the cause of all the bruising in meatworks cattle. None of his team had horns. He owned an excellent polled Cherokee Brahman bull and by using him over Murray Grey cows had produced the world's first Murray Greybra, Susie. By the same bull, out of black Angus mothers, were Brangus heifers, both called Susie, as was his veteran Poll Hereford cow. It was confusing at first but the advantages were soon more evident. The cows probably didn't know the difference and 'Susie' was much chummier than 'the old poley cow' or 'the black heifer'.

Working on the theory that white was a colour far better suited to the Tropics than dark cherry red, Pop mated his bull to a polled white shorthorn cow to produce a more tropical Droughtmaster. This breed was derived mainly from the commoner red shorthorn cross or cherry-coloured Santas and red was the colour set in their standards. They weren't interested in Pop's unique white Droughtmasters and an impasse being reached as far as society membership was concerned, Pop switched his attention elsewhere.

We had a day or two before judging to settle the cattle in. Stalls were mucked out at daylight, the dirty straw and manure heaped in piles for removal in the alley outside the stalls. Cattle were fed, watered either by bucket or at a communal trough, groomed and exercised.

This sounds simple but Cairns water was fluorinated and bulls from the bush don't like fluorine. They wouldn't drink. Being bone-dry and unable to spit, they wouldn't eat. We worried as their flanks grew hollower and they took on a dejected look. Pop Westman took pity on us.

"Here," he said, offering us something black in a tin, "molasses. Put it in their water. They can't taste the difference then."

You can't beat experience. They slurped up their molasses-flavoured water as if it came straight from the Laura River behind their Crocodile home. Instantly, as their sides rounded, their look of healthy well-being returned.

Grooming was easy. It could be done safely in the stall. Exercise was a different matter. No sooner would we venture timidly out, heading for the small, railed, judging ring than a vehicle would come in noisily through the entry gates with no warning but with a Gallic flourish and dash, completely demoralising both us and our charges. We thought we'd never make it across to the safety of the arena. Eventually, the bulls became more streetwise and barely took any notice of vehicles swishing past at very close quarters and seemed to ignore the thousand other city noises.

We were even able to tie them out on a rail as we watched with blase indifference a man and a woman trying to bring two reluctant brahmans down past the horse stalls to the cattle area. It wasn't terribly exciting. Neither animal wanted to go. There was a good deal of hoozling and an occasional slap of lead-rope on rump until the man's bull got the message. Whoosh!

Onward it came at the thunderous speed of a cowboy's horse with a steamed-up posse or an irate band of Indians behind it. Its handler ran as fast as he could to keep up and not lose the end of the vital lead-rope. Encouraged, the girl's bull followed but with slightly lessened celerity. She was able to keep up with her charge.

There was a water tap near the ring. It was where we filled our water buckets, but the bull-handler wasn't looking for a drink. He was looking for something substantial to loop the rope around to give him better purchase.

It was a bad decision. He had barely wound the rope around the stand-pipe when the bull gave a lunge. The pipe snapped off at ground level and went flying, tap and all, over his head. We didn't see where it landed. Our attention was riveted on the spectacular aquatic display the broken pipe generated. It was magnificent even if it did drench the lot of us.

Our bulls were reassured that it was nothing to be afraid of. Someone with quick wits flung open a gate in the fat cattle yards into which the fleeing bulls sped with alacrity and someone else ran to the Secretary's Office to report the broken pipe. Pop Westman's voice restored the normality.

"Looks like Cairns is acting true to form, eh, Mrs Wallace?" he drawled.

I didn't like to comment. There, but for the Grace of God.......

Judging went off as we'd expected. Mr. Atkinson's bull and cow won the Brahman awards. Monty Atkinson, one of the pioneers of the Droughtmaster breed was judge. Grub and Wonderboy each won a blue ribbon in their class. No one else was against them so the victory was a hollow one.

The bulls behaved well though, until the moment when the ribbons were presented. They hadn't encountered anything like that slithery, snake-like thing in their pre-show training, nor was Grub all that keen to have a male steward come close enough to slip the ribbon round his neck. We'd broken him in to lead, to wear a rug to bed, to be reasonably well-behaved and not to be too frightened of blaring city noises and bright lights, but he was totally unprepared for dangly ribbons of felt that tickled and for over-familiar strange men.

We made mental notes for next time. A ribbon-training course would be introduced into the curriculum and men would be invited to handle our charges. The all-women environment was inhibiting.

The children were overjoyed with the two blue ribbons and were taking them back to be displayed along the front of the stalls as was the practice when a steward called, "Champion Cross-bred Bull. Woondoo Wonderboy and Crocodile Dorothy's Pride into the ring please."

That was something we hadn't thought of.

Mrs. Atkinson placed the broad purple sash around Wonderboy's girth. He didn't flinch. She was a woman and he was learning fast about these ribbons. I was embarrassed. It didn't seem right to get a champion ribbon when there was no competition.

"Nothing to worry about," Mrs Atkinson assured me. "Why penalise those who come because of those who don't? Besides, I've seen Monty refuse to award a Champion if he didn't think it was warranted."

That made us feel much better.

A quick lunch for us and the cattle, with a post-mortem of the judging thrown in, a hurried attempt to have the cattle looking their best and it was time for the Parade. A young ringer from the Peninsula had been haunting us all morning. Pinky was down for a break and didn't know what to do with himself in town. Could he lead one of the bulls?

Pop Westman was always short of leaders but I gathered Pinky didn't want to lead an old poley cow adorned with one of Pop's rope halters. He'd much prefer a more macho bull with leather and brass studs and the chance to be close to Janice.

"He could take Wonderboy," Janice suggested.

Pinky was a brawny six-footer and the children had no problem with Wonderboy. Just a tug on the lead-rope and he got the message.

"Righto," I agreed and went to offer my services to Pop as a leader. We had only the three youngest children with us and Bill was to keep an eye on them.

Pop was delighted to see me. He'd already given the Poll Hereford Susie to a honeymoon pair from N.S.W. The bride held the groom's hand while he took the business end of the rope halter. Pop's son and his groom, Pauline, took the Brangus. I could have the World's First Murray Greybra. Her baby calf would follow.

As I took my place in line behind the honeymooners and the Brangus I felt a faint tremor of apprehension as I heard Pop say, "She needs an experienced leader."

Janice and Pinky followed. The recalcitrant Brahman that wrecked the water tap and his mate were way in the lead behind Mr. Atkinson's team headed by his Supreme Champions.

All went well to the holding area just outside the ring. It was a special occasion. The Governor General, on a visit to the North, was viewing the parade from the vantage point of the announcer's box, perched high above the ringside. Fred, from the Department of Primary Industries and one of the honorary cattle stewards, was jotting down information for the announcer. Who was who and what they'd won. The World's First Murray Greybra intrigued him.

"What about the calf? Is it the World's second?"

"No," said Pop's son. "She's by a Brangus bull."

"Oh," said Fred as he turned away and stage-whispered, "World's first Murray Greybra and little Panty Girdle."

The Brangus heifers were feeling fractious. They'd done the shows from Rockhampton up and were sick of Grand Parades, being tied up or led around. They'd be happier back home in their paddock again. They demonstrated their feeling by fidgeting and playing on their leads. A good jerk from Pauline and her heifer remembered its manners while its companion sulked at the thought of a good game thwarted.

The ring was more than usually adorned in honour of the Governor General's visit. Potted palms marked the route and had to be investigated by Wonderboy. Who knows? They might be edible. Pinky didn't seem to be concerned as Wonderboy veered over to take a bite.

The cattle moved majestically clockwise around the ring, passing under the Governor General's stand while Fred's commentary droned on. Suddenly there was a scuffle in the ranks of the Brahmans. They were just passing the suspect amplifier and the two rebellious Brahmans were putting on a show of their own. The one led by the girl broke loose and instead of going softly in the hopes it'd stay with its mate and be subtly recaptured, she flew off in pursuit trying to anchor the fast-disappearing lead by attempting to jump on it.

The effect was chaotic. The man's bull immediately pulled loose and the two liberated animals played chasey in and out of the parade. This suited the Brangus heifers who began to corroboree, squealing with delight.

Ever inquisitive, Wonderboy took a bound in their direction and to our horror, Pinky let him go! Janice had Grub in low-range, first gear, on a very short, tight lead. The World's First Murray Greybra was wrestling with me to get her shoulder past my restraining hip bone so she could join the heifers. The old Poll Hereford, veteran of a score of tumultuous parades, took advantage in the melee to stop to crop the grass. Her leaders walked on hand in hand oblivious to the fact that they'd pulled the rope halter off and were just leading an empty headstall as it trailed behind them over the grass. Their cow had got off the line.

Fred was in the ring, maneuvering skilfully to haze the girl's bull into a huddle with the Droughtmasters and Devons who proceeded regally as if nothing were amiss. Fearing some zealot might shoot my Wonderboy as a mad escaped bull, I hissed to Fred as he passed, "Will Wonderboy be O.K.?"

"Don't worry," he assured me. "The G.G's enjoying it immensely."

Seeing Fred had almost accomplished his mission in cornering the initial troublemaker, the scarlet-coated Ring Steward, mounted on his thoroughbred, decided to do a bit of campdrafting and get the other bull. Fortunately, by this Pinky had re-captured Wonderboy who'd met up with the chastened heifers.

The Ring Steward reined up his horse and, crouching over its neck, galloped in the direction of the second bull. Momentarily everything was quiet. Even the sideshow music seemed to cease its incessant blare. The cattle were so shocked they froze for the moment, too.

"Pull off, you stupid bastard!" Fred's voice rang clearly around the arena.

The steward stopped in a flurry of dust and torn grass and the uproar renewed. It was each man for himself as we clung to our lead-ropes and indirectly to our charges. Little Panty Girdle hung as close as a tick to her mother's side and the newlyweds noticing their cow was gone, hurriedly re-caught her.

One circuit was enough. The Governor-General could enjoy someone else's Parade. The Brahmans led the way out, past the circling dairy-cows and horses and towards the long alley that led to home and safety.

From nowhere the second miscreant Brahman appeared, his owner clinging to the rope but not hampering the bull's escape in any way. A middle-aged woman coming in the opposite direction was nearly mowed down as they rocketed past. The woman had a sense of humour. She laughed as she retrieved her hat and bag from where she'd jettisoned them.

Pop Westman appeared from nowhere to direct our strategic retreat, rally the troops. "Keep the cattle going," he advised in confident tones. "Just keep them moving."

I was having trouble keeping mine from moving quite so fast. I was sure my hip would be bruised black from trying to jam her with it and slow her down. There was a cry from the leaders as the escaping bull swung sharply around a break in the stalls, slamming the man at the end of the rope into the back of a parked truck. He got up once and fell. Stood again and crumpled.

"Keep the cattle going," intoned Pop in the most soothing of tones. "The Ambulance're coming."

Sure enough they were, running as fast or faster than the bull in their efforts to get to the prostrate man. We heard later Kevin had broken a rib but he was back at the show that evening apparently not too much the worse for wear.

The bull, rather dazzled by the havoc he'd caused, surrendered and was led away in disgrace at the end of the line and put in a yard. What happened to the girl's bull was a mystery until we got to the stallion box at the end of the horsestalls, just before we crossed the exit road to the cattle section. A rather ruffled Fred, his usually dapper Don Ameche moustache twitching, was standing in the doorway with a rope in his hand.

"I feel an awful fool," he stammered. " I've got him — but I can't get him to come out."

The bull was hiding, a nervous wreck by this, cowering in the comforting blackness of the windowless stall. Can't say I blamed him. He just couldn't cope with the fast-lane of city life.

Pop Westman stopped and peered in.

"I can't get him to come out," repeated our rather abashed D.P.I. man, the lead slack in his hands.

Pop nodded and Janice and Pinky led their charges in. Reassured by their presence the renegade emerged shyly and was safely secured with his mate in the cattle yards no doubt vowing never, under any circumstances, to come to town again.

With deep relief I handed my Susie over to Pauline's care. Wonderboy and Grub were already tucking into their hay and Janice was boiling the billy. As we sipped our tea, settled our nerves and the kids argued over possession of the ribbons, I knew we were hooked. We'd all been bitten by the show bug.

Kevin didn't carry on with show cattle after that attempt, but Janice's confidence was undented. Packing up ready to leave she lovingly folded the purple champion sash before putting it away.

"The first!" she said. "Next year we'll have Deidre, Chris and Leilani....."

Her voice tapered off as she indulged in happy visions of the next year and the next Grand Parade.

I shuddered. But time is a great healer. We were hooked. Properly.

Bully at Cairns Show - Champion Braford and Reserve Champion, All Breeds
Photograph courtesy of Queensland Country Life

Prudence and calf in Cairns, with Laura, Lennie, Billy Kid and Lee

Bubba at Cairns Show, 1970 - Champion Braford female and Reserve Champion, All Breeds. Photograph courtesty of Queensland Country Life.

Virginia with Nancy and Johnnie at Daintree.

Nancy and Roysey's mare. Photograph courtesy of Cor Jacoby.

6

A Recurrent Fever

There's no doubt that when the show bug bites, recovery is slow and resistant to good advice and commonsense. We became addicts.

Wonderboy and Grub went their separate ways, Wonderboy to his own herd of concubines at Crocodile and Grub was sold to Stan Watkin at Kalpower. His was my first cattle sale, thirty pounds, the equivalent of sixty dollars today but quite enough to splurge on toys for the children, a luxury they didn't often get, and an R.M.Williams woollen swag rug for Bill. He still has it, nowhere nearly as thick and fluffy, but still usable.

Janice wasn't quite right when she said we had Deidre, Leilani and Chris for the next year's show. We ended up with only two of them. They were all milkers' calves. Chris's mother was Christine given to me by Louis Fischer, a very stylish cherry-red cow with a hump and Zebu characteristics. The red was a legacy from her imported Santa Gertrudis sire and the rest were handed down with variations from the remote Melbourne Zoo ancestor. Chris was a real pet and attention seeker.

The two heifers may have looked very much alike but temperamentally they were poles apart. Deidre was one of Dorothy's granddaughters and had all the placidity that came with that family. Leilani had more style. She was quite an eye-catcher. The trouble was that she had a liberated mind. She wasn't bitten by the show bug and had no faith in our judgment on any decision that entailed her being paraded in a halter before a crowd of leering humans like so much beef for sale.

Chris and Deidre, also known as Peanut Paste because of her similarity in colour to that product, were happy to do anything we asked of them. They lapped up the attention and haute cuisine diet. This was the life for them. Teaching Leilani to lead resulted in a complete role reversal. She led us a merry dance and Janice, in particular, a spectacular rodeo-like round of the yard, ending with Leilani delivering Janice a none-too-gentle butt up the ribs. Definitely a 'cross' animal when upset and not a Grand Parade prospect. With little reluctance we crossed her name off our list.

Deidre and Chris were still crossbreds. They were third cross or seven-eighths but the goal posts had changed. Another cross was now necessary for purebred status. That didn't worry us unduly. They were an attractive pair (Chris was the colour of peanut paste, too) and we had purebred calves to look forward to for next year.

We were determined not to make the same mistakes with this pair as we'd done with Grub and Wonderboy. Any male visitor, including the Presbyterian Moderator up from Canberra, was invited to step up close and pat our show poddies. The latter enjoyed it. I can't speak for the Moderator but he didn't object and politely carried out his public relations exercise. Ribbon practice was on the daily training menu and a transistor blaring at maximum volume was supposed to accustom our charges to sophisticated non-rural noises.

The children played a part in quietening the cattle but the time had come for John and Nancy to board away. They went to the students' hostel, Woodleigh College, in Herberton. Some of their friends were already there. For a short time Billy Kid stayed with his old friend Lucy in Mossman and attended the State School there. It was certainly very quiet with only Laura and Lee at home but with just two children to teach it gave us more time to spend on our show poddies' formal education.

Once we decided not to spend any more time on Leilani, she contrarily took an interest in Chris and Peanut Paste's training, watching closely and observing with a jealous eye the extra rations they were given. She became much more tractable but with the Grand Parade constantly in mind, we easily resisted any fleeting inclination to give her a second chance.

Leilani's mother was a special favourite of Bill's, a big yellow Zebu cow that was a most spectacular milker. In the first few weeks after calving, her baby had no hope of drinking anywhere near the milk her factory churned out and we had to milk her night and morning to prevent udder damage, a routine Bill adhered to very strictly.

While I was away having Laura I received a letter from Bill. The opening line sent a warm wash of emotion flooding my veins and brought a fond lump to my throat.

"Dear Mate," he wrote, "I wish you were here."

I paused a little as I wiped a sentimental tear from my eye and read on.

"Lady is due to calve any day and I have to meet George on the boundary. If you were here I wouldn't have to worry about her."

So much for visions of romance but it was still nice to be appreciated for my more practical points.

With only the two animals it worked out well. We took ourselves to Cairns in style with Chris and Peanut Paste in a homemade crate of tea-tree rails on the back of the Land Rover. For a while we had problems as, unused to the sophistication of motorised transport, Peanut Paste clung desperately to Chris's side for emotional and physical support. Land Rovers weren't designed to carry the weight on one side over one wheel and the screeching noise as the back of the tray descended onto the revolving tyre was quite unnerving. To humans as well.

We'd taken a spare rail with us for emergencies, together with the usual axe, wire and pliers. You never knew when they might be needed and it didn't pay to go unprepared. Going down before the show we probably wouldn't have met another vehicle until at least Carbine. Lonely roads are good self-sufficiency training grounds.

Janice and I used the rail to divide the Land Rover tray longitudinally into two thus splitting the weight evenly over both sides and had no more trouble.

We camped again in the feed room. Experienced old-stagers this time, I went to the Secretary's office as soon as we unloaded to get a replacement light bulb, while Janice and the kids unpacked and made tea. Someone must've had a good market for second-hand light bulbs as, soon after, all the ones from the cattle stalls disappeared as well as the one from the feed room. Not wanting to walk to the other end of the grounds again to see the Secretary, we simply took the ones left in the horse stalls. First in — best lit.

The number of led cattle was increasing but we were the only cattle people camping on the grounds that night. This was fortunate for the others if not for us, as a pack of louts came rampaging down. They ripped out the phone that connected the cattle section to the Secretary's office, removed the light bulbs we'd thoughtfully borrowed from the horse stalls, dropped the rails on some of the stalls and tried to ride the sleepy cattle and chase them out.

It was unfortunate that our shut door challenged them. They stormed in only to be confronted by Janice with a hay fork, prongs-up at the ready, and me behind her with the lump of tramline we carried as an anvil for running repairs. I think they just meant to collect another light bulb but their shock at finding armed Amazons was at least equal to ours at being broken-in upon. They fled. Trapped in the tiny room, we couldn't.

A Brahman heifer belonging to Eddie Hackman was preparing to follow them out the gate when we caught her, put her back, made sure everything was secure for the night and hopefully crossed our fingers.

Even then we didn't sleep soundly, though the children, tired by their trip, never woke. With the door barricaded with bags of feed, hay bales and whatever we could find,

Chris and Peanut Paste make a lopsided load for the Land Rover.

we were grateful none of the intruders dropped a match into the straw bedding. We'd never have got out.

One of the studs who ventured North that year after showing the Central circuit, was Cherokee with Mr. D. (Delandelles) and his daughter Elsie in charge. They stayed at a nearby motel and we were thrilled and flattered to be asked to keep an eye on their team during the night. Yabba, one of their grey females, was a superstar as far as we were concerned. Our ideas of heaven were speedily amended. As well as having perpetually green grass, running water and white railings we also added a Yabba or two in the front paddock, preferably with baby Yabbas at foot. The others in the team including the dark grey bull Repucho were equally as good but Yabba had that extra personality and genuine animal magnetism that set her apart.

We watched in awe as Elsie exercised her on a loose lead held daintily between two fingers with the little one crooked at an elegant tea-party angle. Yabba played to the crowd cavorting and gamboling like a lamb crossed with an elephant without causing any tightening of that slack leadrope. She knew she had what it took and seemed delighted in dazzling the local yokels.

Like Mr. Atkinson and Mr. de Tournoeur, Mr. D was generous with his advice and did all he could to encourage us, never seeming too busy to join us for a cuppa. He gave us the recipe for the feed he used and which we could mix ourselves. A bit more sophisticated than the bag of pollard to one of laying mash formula. He introduced us, both cattle and humans, to the delights of boiled barley and successfully demonstrated how to stand a beast correctly for the photographer.

There is an art in this, both in the posing and the photo-taking. John Boydell was a master at the latter and Elsie at the former, but for some reason Yabba didn't seem to have her mind on the job. She stood, almost slumped, like a melting jelly, extremely relaxed and certainly not projecting her magical superstar image. Various bystanders and underlings tried the usual gamut of attention-getters - waving arms, stamping feet, clicking fingers, imitating the call of a lovesick bull. All failed to impress. Yabba looked bored stiff, disapproving and inattentive.

Mr. D. spotted a woman wheeling a toddler in a stroller through the gate and after a few moments of charming diplomacy, the woman removed the baby and gave Mr. D. the stroller.

"Ready?" he asked John and Elsie. They were, so he wheeled the stroller at the slumbersome heifer.

She didn't move physically but her attitude changed in the speed of a lightning bolt to assume automatically the required head-up, shoulders back, ears-pricked, super-intelligent stance.

John took a perfect picture.

"She's never seen an empty stroller before," explained Mr. D. So far we haven't needed to use that useful tactic but the knowledge of its success is with us just in case.

We were becoming more seasoned and making fewer mistakes. Not only did Chris and Peanut Paste win their crossbred classes (and championships) but Peanut Paste went on to win Champion Districtbred Female beating a local purebred. As a stud sire, Chris's missing eighth went against him in the line-up but he certainly wasn't disgraced.

1964 was memorable for other reasons than Peanut Paste's handsome purple sash. Bill's brother Hardy turned up unexpectedly at the show. He looked very important, smug and secretive - all quite normal - but we did wonder why he'd come down from Merluna for the Show when we knew he'd been very busy at home and much preferred race meetings to the relative dullness of shows.

The Boss, Bill and Hardy's father, had financed Hardy and a partner into Merluna and its Siamese twin, York Downs. By degrees Hardy had bought out both his father and his partner, though it left him with no money left for luxuries like additional fences and paddocks.

Merluna and York Downs had been separate runs in the early days but not long after World War One they'd been put together as a state-owned station by the then Queensland Labor Government. It wasn't an attempt to socialise a rather independent-minded industry but a 'courageous' venture to provide cheaper meat for Queenslanders. The success of the scheme was debatable. Government mines and stations made extravagant losses but I believe other ventures, like the State pub near Babinda, were much more successful.

Together the two runs comprised a large area of grazing land, a size not unusual for the lighter-carrying Peninsula but one very difficult to tame without barbed wire. Even today many of the far northern boundaries are unfenced. Neighbours muster together each claiming his or her own cattle and sharing the unbranded 'cleanskins' that can't be 'mothered' - attributed to a branded cow.

One of Hardy's neighbours was newly arrived after selling a business on the Townsville to Mt. Isa rail line. He had the wherewithal to build the fences and paddocks which soon took shape over his run. His cattle numbers weren't as generous, so he engaged contract musterers to increase the branded herd. Not surprisingly he wasn't popular with the older hands who may, of course, have been envious of his affluence. Wild stories began to circulate of branded cows with unbranded calves being mustered off their home runs, the cows shot, the calves branded with the new owner's brand, secured behind the barb wire and left to take their chances on survival.

A certain respect was paid to poddy dodgers who risked life and limb to capture unbranded bulls, noted neither for their tractability nor for their innate love of man, on the no-man's land of the boundaries, but shooting branded cows to get the calves was considered the utmost depravity.

Hardy heard a rumour that this new neighbour was planning to go to the unoccupied York Downs, take Hardy's horses that he kept there for mustering and use them to muster York Downs cattle at a face back to his own run. The man to be in charge was an old mate of Hardy's who'd worked with him and also for the Boss at Butcher Hill. Ruth often talked of his enormous strength and told how Texas, as she called him, pushed the others aside when they were branding foals, picked the foal up bodily by himself and smoothly lay it down again, right side up, ready for branding. An amazing feat.

Hardy's quirky sense of humour often got him into trouble. He'd paid the fines of two men caught duffing his cattle because they'd once been his mates and/or had worked for him. The police in Coen had a letter of complaint stating that Hardy was intimidating the newcomer's aboriginal stockmen. He'd told them he had a gun that could shoot around trees and with an accurate range to infinity could hit anyone caught pinching his cattle. Automatically hone in on them and - kazoom!

While he wouldn't confirm or deny making this claim he did little to stop it. He was the owner of one of the first two-way radios in the top part of the Peninsula and gleefully said the story of the preposterous gun was helped along considerably when he talked to people on his transceiver and they answered back, in the middle of nowhere, with no human forms to back up the voices.

He'd had some contact with the contract musterers as well as paying the fines. One came to him in an extremely irate state of mind. He'd mustered Hardy's northern boundary without Hardy's permission but with promises of rich rewards from the neighbour for unbranded cattle of all sizes, ages and sex with no questions asked. He had speedily mustered them, only to find the budget for buying cheap cattle had run out. The deal was off. Too many purchases and not enough sales. The money bin needed replenishing.

But so did his. He failed to see the humour of the situation but was a quick thinker. Furious at being double-crossed he rode off to find Hardy.

"Do you want 'em?" he asked. A bit unnecessary seeing they already were Hardy's. "We tailed 'em out for a week along the river. They shouldn't have gone too far."

Hardy's money supply also needed replenishing and cattle running so far from the main station were obviously at risk. They came to a mutual settlement, Hardy got on his trusty two-way to arrange a sale and transport, and within days the entire mob, bullocks, bulls, cows and calves were aboard the cattle barge at Arukun en route to Cairns. Arukun had been mustering too and were happy to make up the boatload.

We'd often heard strange radiograms sent over the Aerial Ambulance radio service. I'd written one down hoping to find out what was behind it but Hardy would only smile and look secretive. 'Have sent men contact your plant to pick up my cattle. Stop. Unnecessary your plant burn or muster Moonlight or Leichardt', the signature a formal 'Wallace'. Burning was a favourite practice to coax cattle onto a certain area to make use of the green pick that followed the burn-off. Practiced near an unfenced boundary it enhanced the value of the procedure by luring cattle across the proprietarial line. Moonlight and Leichardt were on Hardy's property.

Perhaps Hardy's sense of humour didn't serve him well when he sent another radiogram in reply to the bush telegraph communication concerning the York Downs horses.

'Hear interested York Downs horses. Stop. Can inspect same ten pounds per head. Wallace.'

It wasn't well received. As anyone on the network with a set running was at liberty to listen to everyone else's 'traffic' the person in receipt of the wire wasn't amused. Hardy should have known better.

Instead he let it be generally known that he was off to the Cairns Show for a break. What he didn't broadcast was that he'd left Vera and John Harris at York Downs.

The water supply at York Downs was barely adequate for household purposes so the horse plant was turned-out on the lagoons at Myall Creek nearby.

As expected, the uninvited guests arrived and were mustering (using their own horses) when John found them. After some heated words on both sides John retrieved the York Downs cattle and took them back to the outstation. Next day, the musterers returned and again John met them. This time things didn't go as smoothly. After a spate of choleric and slanderous name-calling, shots were fired. Texas had a fifteen shot Winchester .22 rifle and John the usual .38 revolver cattlemen carried to deal with intractable cleanskin bulls.

The aboriginal stockmen wisely decided they had urgent business elsewhere and left them to it.

Shots flew frenziedly. John stopped five bullets, including one that he still carries lodged against his spine. Texas scored three. One in the arm and two in the 'guts'.

Vera, who quickly assessed the situation, drove John to Weipa from where he was flown to Cairns. It was at this point that Hardy received the news. His smug self-satisfaction vanished in the flick of an eyelid. Things certainly hadn't gone the way he'd expected. Chartering a plane he returned and flew Vera down to be with John.

Meanwhile Texas headed for home but, bleeding heavily, he became 'slewed', disorientated, and got off his horse to try to rest and regain his senses. He was found later and evacuated to Cairns Base Hospital where he spent a couple of weeks in the bed next to his fellow gunfighter.

Neither laid a charge against the other so the police had no cause to proceed with inquiries.

Perhaps the guns went off accidentally while they were cleaning them. In any case things were a bit more peaceful in the fartherest North after that.

* * *

Brahmans were becoming popular and when we sold Chris - to a very good home, the A.A.Co - at the Mareeba Bull Sales at the end of the year we got $1000 for him, a big increase on Grub's sale price. We thought it was pretty good even though Mr. D. said, probably with truth, that with his brand on him Chris would have brought four times as much.

Deidre had a less happy future. We should have given her another name. Deidre, we were told, means lady of sorrows. Butcher Hill was sold and the new owner was using the yards and musterers' hut behind our house, all on Butcher Hill country. Not realising the danger, he didn't shut-up the hut when he left. It was kept closed as a drum of arsenic dip had rusted and leaked onto the ground years before. Arsenic seems to attract cattle.

It attracted the luckless Peanut Paste when she saw the open walkway. We searched for her all day looking everywhere but in the usually out-of-bounds hut. Bill finally made the discovery and it took some time before we conceded that he wasn't fooling and faced up to the sad truth.

Deidre was dead.

* * *

In reality, the show poddies took up very little of the year, about two to three months of preparation, and once they were going well, their needs could be attended to in a couple of hours each day.

There was enough to keep us busy without them. Bread had to be baked daily, but once the Cooktown to Laura railway closed and the road improved sufficiently, George Watkin from the Laura pub ran his truck to Cairns every week - weather permitting - and we were able to get fresh fruit and what veges we couldn't grow, as well as the odd loaf of 'baker's' bread, up with him. He also brought our mail from Laura. Progress was certainly well on the way.

Correspondence school lessons still had to be supervised and took up a sizeable amount of time, for if a Home Supervisor were tempted to leave the student get on with it while she did something else out of sight, the resulting schoolwork often left much to be desired. It wasn't unusual to be greeted by a blank page or unsolicited graffiti, sketches of bulls and horses and other creative outgoings not calculated to gladden a Brisbane-based teacher's heart.

Billy Kid began school the year Janice came to work at Crocodile. She taught him while I supervised John and Nancy, thankfully by then, in the one grade. Laura started next and two years later it was Lee's turn but by then the older ones had gone on to State school, boarding at the hostel in Herberton. We adhered, more or less, to school hours but holidays were governed less by Saturdays and Sundays than by days when we thought the activity planned offered a better alternative to sitting in front of schoolbooks.

Bill usually had a permanent 'man' of whatever colour and a young Daintree lad to help. The latter were part-aboriginal sent up to us with Louis' recommendations and, unlike Bill's droving 'boys' from Hopevale Mission whose wages, less pocket-money, were paid in trust to the Hopevale superintendent, they were not 'under the Act'. They had the same rights as the majority of Australians and their wages were paid direct to them.

There was little chance of the young fifteen- or sixteen-year olds spending much except at Laura racetime, so at Christmas they went proudly home with a handy nest-egg. One young lad, Harry Burchill, was a special favourite. He was a son of the Billy Burchill who had been so kind and understanding to me when Laura was small. His brother, Billy, also worked for us for some time. Both excellent workers, Billy was very shy and introverted - Harry the opposite. It was impossible to be ill-humoured with Harry around, he exuded good-will and was at all times eager to please. Bill had great hopes for him, showed him how to spey and taught him station skills that normally didn't interest the other youngsters. Harry was a quick learner and took a gleeful pride in his accomplishments boasting, not without substance, he'd be headstockman one day. That day didn't come. The heady delights of town intoxicated him in more ways than one and he gave the bush life away. He lived it up for a brief period then, in a fit of despondency, took his own life. It was inconceivable, but Harry, the ever-cheerful extrovert, was no more.

At Butcher Hill, Paddy was 'under the Act' and his money was held in trust for him by the local Protector of Aborigines, the Cooktown police sergeant. Paddy's wage was the same as Bill's when he worked for his father, seven pounds (fourteen dollars) a week. I 'd been paid four pounds but Leo, Paddy's wife received no set wage. Paddy was given a pound a week pocket-money and the rest, less the price of any clothes, toiletries etc he and Leo ordered from town by way of the Boss, was credited to his account.

Plug tobacco, a solid tobacco in a small cake or plug, the heavier Champion or the smaller plug, Sunlight, was rationed out each Sunday before the weekly walkabout and Paddy seemed to make his last. I don't think Paddy's second wife, Mary, smoked but Leo, who shaved the sides off the plug with a pocketknife to stoke her pipe, seemed to run out towards the end of the week and often prevailed on Ruth or me to filch another from the storeroom at an appropriate moment when no-one was likely to be guarding the tobacco supply. Her pipe's appetite was insatiable and her withdrawal symptoms severe.

As Paddy had been working for wages - albeit small ones - for almost all his life he'd accumulated a substantial sum. At one stage Ruth was day-dreaming about buying the Laura pub. Paddy offered her his savings, then in the four figures, towards the deal but fortunately it didn't get past the day-dream stage. Ruth was rather soft-hearted and too many drinks and meals would have been 'on the house' for good profit-making business principles.

Before Butcher Hill was sold Paddy did withdraw a hefty part of his savings to buy a brand-new Land Rover. With his new wife Mary and the two boys Peter and Michael, he went walkabout the modern way each Sunday. The purchase would no doubt have been supported by the Boss who conveniently gave Paddy rocksalt to take out with him to top-up the salt-troughs on his Sunday outing. Though they took with them a supply of 'tea leaf' and sandwiches these were supplemented with whatever tasty morsels the walkabout would yield. Turtles and fat little perch from the basalt springs were eagerly sought and cooked on the coals.

Peter and Michael were family names. Paddy's older brother who died, possibly of rheumatic fever, was a Peter as was Hardy's son so that there was a White Peter and a Black Peter. There was a Michael in each family as well but to add to the Michael confusion, Paddy had two brothers, both Michaels.

He came from a tribal family. One Michael was a full-brother the other a half-brother or as it was worded, 'same father different mother'. The full-brother had taken a different surname from his employer, Michael Webb, but the half-brother was also a Wallace - Big Michael. He worked for Jack Wallace outside Mareeba, at the slaughter yards and as a drover bringing in the drafts of cattle Jack bought for the shop. Big Michael was a better than average camp-cook and could turn out delicious currant brownies and apple tarts in a campoven. Paddy's sister Emily was talented in that way, too, and made the best Johnnycakes (scones cooked on coals) I've ever tasted.

With this background of family names it sent the hairs on the back of my neck prickling years later, after we'd left the Peninsula, to hear on the radio one morning the impersonal news item,' Another black death in custody .....at Wujil Wujil ....Michael Wallace.'

Paddy's wife Mary came from Wujil Wujil.

As soon as I could, I contacted Bill's sister Joyce on the RFDS radio and yes, it was Paddy's son Michael.

I had always assumed Paddy, Peter and Emily had been 'taken away' from their family by the authorities and given to Bill's father to rear as workers, but Paddy often spoke of his mother and father and kept in sporadic touch at racetimes and other such occasions, with the two Michaels. That didn't correspond with the usual procedure when children lost contact with their family completely.

A few years ago I heard the story. Bill's grandmother had been a young child when her father took-up Butcher Hill in the 1870s. He was one of those more enlightened men who used aboriginal labour and 'let in' the aborigines to live in comparative safety around the homestead. His daughter Isabelle learnt the 'language' almost as soon as she had mastered her own and continued her interest in the original inhabitants when she married and went to live on the Endeavour River north of Cooktown. Here she came into contact with Maggie, mother of Paddy, Peter and Emily and her husband Billy.

In an extreme drought year Maggie came in to Glenrock from Flaggy where they were living in a tribal situation - with an annual handout of blankets and a tomahawk or two but little else. Her manner was agitated and when Isabelle asked her what was wrong she told her there was no tucker for the children. Could Isabelle's son, Charlie, take the two boys? He did and the boys went out to Butcher Hill.

However there was no relief from the drought and a haggard Maggie came in again to see if Charlie could take Emily too. Isabelle wasn't too sure, though she would have taken the girl herself to ease Maggie's mind. Charlie did agree and Joyce remembers vividly her father arriving home from Cooktown with the tiny and totally unexpected Emily perched up behind him on his saddle horse. It must have been a long uncomfortable ride for them both, taking the best part of two days to cover the distance by horse. Contact was maintained with the family until Maggie's death and even after Billy died.

When Peter was about sixteen or seventeen he was struck down with a fever and became seriously ill. Some of the others were sick also but Peter was by far the worst. The plant had gone out camping leaving Peter at home with Bill's mother and an old retainer, Black Charlie (not to be confused with Bill's father, another Charlie).

Peter had been getting steadily worse, but one evening when they went to attend to him, Bill's mother said he looked so much better she was relieved, thinking he'd passed the crisis and was on the mend. However her anxiety was rekindled when he told her he could hear lovely music and see a beautiful light and that his father was there beckoning to him.

"Mumma," he told her, "my father wants me to go with him."

She thought he was hallucinating but as he was smiling and looked very much brighter, she stifled her concern.

Later that night, when she and Old Charlie came back to check on their patient, Peter was dead.

As they were on their own at the homestead with no other help available, they buried the young man, wrapping him in his blanket with his whip and spurs beside him. When the musterers returned someone was despatched to Cooktown to inform the Protector of Peter's death and burial.

"Bit of a coincidence," the sergeant remarked. "His father, Billy, died just the day before him."

No one, of course had known. Although no postmortem was done it was thought Peter died of rheumatic fever, a disease that was prevalent in the area at the time.

Butcher Hill was sold at about the time when Protectors were outmoded, aborigines obtained full citizenship rights and stockmen like Paddy had to be paid the award wage.

Bill's father did consider it but rejected the idea of taking Paddy and his family with him to a small retirement block near Mareeba. Paddy, also, was loathe to leave his home and wanted to come to us.

There was no way we could, at that time, afford to pay Paddy the award, build the now much higher standard of accommodation for his family and look after them as well. Paddy wanted to come to us on the old terms but ' exploitation' was a favourite word and as newcomers passed through the area it was used often. Cattlemen were considered white oppressors. Reluctantly Bill declined Paddy's offer and we were filled with remorse years later when Bill's niece Mary related to us that Paddy thought 'no-one wanted him.'

Paddy stopped on at Butcher Hill with the new owner for a while but his reputation was well known in Cooktown and he was offered a job at the slaughter yards there. Both Michaels had done similar work. Unfortunately town living got the better of Paddy. He 'got on the grog', lost his Land Rover, his money, his family and his self-respect.

We'd moved to the northwest downs country south of Hughenden when we heard of his plight. We were now in a slightly better position financially and wondered if he'd come out there to us. It was a long way from his home. Barragunda, like most properties in the area, had a married man's cottage - not as necessary now with Mary living back at Wujil Wujil. On the grapevine we'd heard Paddy'd had heart trouble, been fitted with a pace-maker, given up the grog and wanted to get back to the bush again. It'd be a long way from his aboriginal friends and relations but we wondered if we could persuade him to come out to us.

By the time we found out where he was it was too late. He'd made a break from the grog and got a job caretaking at Lakefield where he and Bill had both worked, a place dear to the hearts of everyone who ever worked in the cattlecamps there. On trying to contact him there we found he'd just died.

One small consolation. At least he didn't live long enough to know of his sons Michael's unhappy end.

7

Treasure Lost and Found

Our meeting with Mr D. led on to other things. A visitor to Crocodile had put the idea into Janice's head of travelling the shows with a team of show jumpers. She told Janice she'd contact the horse-string's owner and recommend her. If Janice got the job they'd start early in the year on the Darling Downs and New South Wales border, progressing through to the Brisbane Exhibition in August. Janice was floating on a cloud of rosy dreams at the prospect.

Mr. D. didn't think it was a good idea. It'd be all right, he conceded, if she did a few local shows each year, but travelling for the greater part of the year was a different story. I was inclined to agree but Janice was so enthusiastic in her anticipation of the great adventures ahead, that tactfully, I hope, I remained neutral.

"Send her down to me. I'll talk her out of it. Give her some tips on showing Brahmans and send her back."

Mr. D's words, calculated to soothe, unsettled me never-the-less. I knew that once Janice got to the heady atmosphere of Cherokee, all thoughts of return would vanish. Bill and I had already driven down there, our first trip away together for over fifteen years. We'd left Janice holding the fort at Crocodile. The Delandelles family made us feel very welcome and showed us around.

There was certainly a lot to see — the Brahmans, daughter Alison's Bantengs, (tiny Bos Indicus cattle from our northern neighbours), the Arab horses, the massive Percherons and almost all the birds pictured in the book of birds. Mrs. D. had a vast aviary of pheasants, the colours of their plumage and their mode of feathery dress very avant garde. Peacocks, blue, green and lacy white strolled the lawns and picnicked amongst the feeding cattle. A drive along the dazzling coastline finished our first day perfectly with the reflected rays of the setting sun turning the ocean's surface to opal.

But we hadn't seen the real object of our visit - the sale bulls.

We broached the subject after dinner to be told by a surprised and apologetic Mr. D. that he hadn't realised we were prospective buyers. He thought we were merely taking advantage of his open invitation given at Cairns to visit Cherokee and see the sights.

The misunderstanding, if you could use that word, was soon cleared. The next morning we were whisked out to view the bulls. Selecting them wasn't easy as there were hundreds to pick from and so many we would have liked to have taken home with us. Naturally, with the elite sires, the price-tags of our first choice didn't suit our budget, but we selected two very good bulls - a red and a grey. Their names were taken from the T.V. program that night. 'Saratoga' (Trunk) and 'Minnatonka' of 'by the waters of' ilk.

The television was an unfamiliar treat for us and was still a novelty at Cherokee so that the seven of us lined up for dinner along one side of the enormous glass-topped table facing the screen on the other side of the room. It was the adult equivalent to a child's visit to Disneyland with all the wonders of the world around us. With a slightly sinking heart I knew that Janice, once there, would be entranced by the magic and never return. Nor would we want her to. Cherokee had infinitely so much more to offer.

At the end of the year Janice left, bound first for home and then for her stint at Cherokee.

"I'll be back," she promised, blithely unaware of the enchanting temptations of the paradise awaiting her.

"Cherokee'll be the rock you get stuck on," I predicted and was right. Janice stayed with the D's for many years.

Missing her dreadfully, I saw an ad. in the Country Life from a girl wanting a job with cattle and horses. Jillarooing was then a relatively innovative job for girls and young women wanting jobs in the bush were rarities. I followed the omen and replied as soon as I could.

Within a couple of brief weeks I was driving to Daintree to meet Marie who'd come from Caboolture to Cairns. A willing friend drove her to our Daintree rendezvous. First impressions were mutually satisfactory and we drove back to Crocodile together the following day.

Janice was a 'Junoesque' blonde, Marie, petite and slightly built with the most beautiful, lustrous, black hair. In temperament they were also dissimilar. Janice was placid, extremely hard to ruffle. Marie's dark eyes flashed and shot lightning bolts at the slightest provocation from humans. With cattle and horses she was patience personified and we got on exceptionally well. Marie was experienced with horses but cattle were new to her. Luckily I could pass on to her the knowledge that Janice and I had acquired the hard way.

We hadn't started on the show poddies and were just doing the uninteresting, routine dry-weather jobs when Bill's sister, Ruth, upset our day's schoolwork with a phone call from Laura township.

"Can't call in. In a hurry. We'll leave a parcel at the mailbox."

As Ruth didn't elaborate, the message in itself wasn't earth-shattering. School proceeded as usual till smoko after which, without much excitement, we went to pick up the parcel. There was no hurry. Those were the days when mail boxes and 'other people's belongings' were sacrosanct. Adventurous travellers hadn't yet begun the regrettable practice of shooting-up mail boxes, signs and anything that did or did not move. The short trip out to the mail box would be a brief but a pleasant outing before school papers were tabled again after smoko.

Then we saw Ruth's 'parcel'! A jam-packed tea case Marie and I could barely lift into the back of the Land Rover and two ancillary chaff bags stuffed to titanic size and as tight as drumheads.

To prevent sticky-beaks undoing everything then and there, we beat a speedy return to the house. There seemed to be enough to stock the shelves of a decent-sized general store. What a treasure trove! More than sufficient for a lifetime of Christmases for all of us - with birthdays, Easters and Mothers' and Fathers' Days thrown in.

There were lengths of material, flannelettes in plain, stripes and checks, bright satin-cottons and a vivid red velveteen which Mike the Frenchman, rubbing his cheek against its silky softness, enthused was just the colour of the red in the Tricolor. He was so overcome he stood at attention, clasping the velvet to his breast and gave a spirited rendition of La Marseillaise.

There were 'snowstorm' ornaments, the kind encased in a glass or in this case, plastic dome where 'snow' floats gently downwards as the bauble is inverted. The suitability of the choice of snow was hard to comprehend as the figurines were an aborigine with spear and a surfer with his board. Perhaps it wasn't snow but drifting dust and sea-spray.

The children, wildly excited, burrowed into the spreading piles. Tiny plastic farmyard animals, cowboys and Indians, legs suitably bowed to accommodate their plastic horses, blow-up plastic giraffes and bears, length upon length of colourful flower wreaths, Barbie dolls clad, most appropriately it turned out, in infinitesimal bikini swimsuits or frayed-legged jeans.

Johnnie tunnelled past these frivolities to unearth an assortment of paint brushes, a screwdriver and a hammer. Marie was trying on shoes — all size 4 — while I ooh- and aah-ed over the elaborately embroidered tablecloths, one each, with its accompanying table napkins, for Nancy, Laura and me. School was completely forgotten and we were still sorting the manna into heaps when the men came in expecting to find lunch.

We'd heard garbled stories on the bush telegraph about tons of loading being jettisoned from a boat stuck on a sandbank in Princess Charlotte Bay but we had to wait for Fred and Ruth's return to get the details. The Shephards and their men were camped-out on a creek near the bay with Miles Morris and his camp, yard-building. They'd worked more quickly than they'd allowed for and ran short of wire. Two aboriginal stockmen were sent back to camp for another coil or two.

They didn't return.

Ruth and Paddy's sister, Emily, were dispatched to find them.

They didn't return.

Alarmed — it was late afternoon by then — the remainder of the work party swung into search party mode. What they found was almost incredible. The whole beach, stretching as far as they could see was covered with everything from floral wreaths and barbie dolls, through cartons of radios (very much saltwater-affected) to heavily-carved furniture.

The first two on the scene hadn't been able to believe their eyes either. As they rode along the shoreline to the camp they saw strange, brightly-coloured objects bobbing towards the beach on the incoming tide. Confirming that both had seen the same apparition, they warily crept forward for a closer examination, horses snorting and not at all happy with the turn of events.

Some wag had inflated the blow-up plastic toys before throwing them overboard and being light as foam they'd arrived first, propelled both by the tide and a strong south-easter, garish orange giraffes with purple spots, lolly pink elephants, emerald green lions with glowing ruby manes, tigers psychedelically striped in the most imaginative, lurid hues.

No one knew much about drugs in those days or they might have suspected the cook of lacing the stew with Gold Tops or Blue Meanies — 'magic mushrooms'.

The two first-comers passed over everything else for the plastic animals. Gathering hundreds of small cows, horses, dogs, chooks, pigs and turkeys, all in bright primary colours with the exception of piglet-pink for the grunters, they stacked them in towering His and His piles and combed the beach eagerly looking for more. Interlocking sections of white plastic yard rails were stashed with them ready to be turned into sets of make-believe stockyards.

Ruth and Emily by-passed the animal-musterers to rescue the Barbie dolls. Brought up with her mother's constant warnings of the effect of too much tropical sun on tender skin, Ruth agonised over the plight of the dolls in their scanty swimsuits. She and Emily gathered as many as they could carry, then gently laid the others to rest in the shade

Bobbing along towards the beach were orange giraffes, lolly-pink elephants and psychedelically striped tigers.

under the straggly beachfront bushes. The surplus jean-clad dolls had to be content, when the available shade ran out, with being settled face-down in the sand.

Their Hers and Hers piles were growing into mountains with floral wreaths and novelties stacked higher and higher, far too many for them to carry.

Fortunately the men, mystified by the disappearance of two more of their companions, arrived in a vehicle just as an infinite row of red balls the size of basketballs came bobbing in on the waves, but before anyone could gather any, the tide turned, the bobbing balls bowed farewell and dipped and dived on the receding wave crests out of reach. They were never seen again.

A council was held when enough committeemen could be persuaded to drag themselves away from the gaudy treasure on the beach. The two stockmen had a pack-horse, fortunately still there, though long-forgotten by the scavengers. Ruth and Emily could take this horse and fill the packbags intended for the wire with the smaller loot. The men would carry the larger spoils, mostly mysterious crates and enigmatic tar-paper bundles, across the beach to the truck. This included an amazing collection of carved furniture - nests of tables, a writing desk, a cocktail cabinet and lesser marvels in the forms of cane-, bamboo- and folding metal- chairs.

Back at the camp, the aboriginal women left to caretake were rapturous with joy. They proposed an immediate second expedition but their immediate curiosity about the crated and papered bundles held them back. Prised open, the crates revealed bolt after bolt of material in forty metre rolls. Flannelette in white, pyjama stripes, florals and cowboy-shirt plaids. Rich brown, blue and emerald velveteen as well as the red, floral satin-cottons in a variety of bright jewel colours and crates of embroidered tablecloths and serviettes. The tablecloths were in two sizes, the smaller about a metre square and the larger, big enough for a fullsize dining table or a single bedspread.

These were pounced on with glee by the happy camp care-takers. Table napkins were appropriated for use as handkerchiefs, small cloths for tea-towels and the bigger ones for sheets. Ruth indignantly interfered, explained the more correct use for them to the women's disbelief and rationed them out. Later, once they'd learnt the commercial value of their largesse, their supply was available for trade and I swapped a leather shoulder bag, an extravagantly brass-studded belt and some dresses for a big tablecloth. I had given mine to Marie. The serviette/handkerchief supply had run out.

Carbide lights burned into the night and the fire was stoked and re-stoked regularly as the contents of the crates were explored and divvied out. Only the things that would float made it to the beach. How the hammers got there was never explained. Perhaps they were but a few sharing a berth with more lightweight stuff.

Wooden chests of green China tea were hastily discarded as "Mouldy!" only to be picked up later by tea connoisseurs in the second wave of treasure-seekers. Children's and adult's tights were likewise scornfully dismissed with a contemptuous "Shrunk!" until the more sophisticated scavengers came onto the scene.

Electrical gear was of course totally ruined and useless. Two hundred litre drums of tung oil rolled in and spoiled cakes of crushed walnuts, the size of a fair-sized tabletop. The feral pigs grew fat on the walnut meat. Laboratory glassware, pipettes, glass-topped jars were uncrated and ingenuity tested to find uses for them. The jars made great condiment canisters and the pipettes, I was told, were just the things to use to get oil into awkward places.

The tung oil was put to use on a one-off scheme by Clara and her pal. Quick and efficient seamstresses, they sewed the velveteen into tentfly-sized rectangles only to find, tightly woven though they were, they weren't waterproof. Tung oil was applied with a convenient wide brush but though the water repellent qualities improved somewhat with this, the brilliant colouring suffered in its application.

Over the next few days before the yards were completed, Ruth and Emily used any excuse possible — 'make sure Clara put the beef on', 'get my gloves' — to visit the camp and the loot. Naturally, Clara and her friend frequented the beach also. There was always plenty of salt beef boiling and damper cooking but the bewitching treasure was a powerful magnet to the four women's iron filings.

With her usual generosity Ruth divided her share. So much for us, so much for Joyce and Fred's sister Thelma, Iris at Merluna. A bolt of white flannelette and a couple of cane bassinets for an expectant mother in Laura and enough plastic animals, gee-gaws and Barbies to satisfy the longings of all her juvenile acquaintances.

Trade and barter were in full swing. I made plaid shirts for the aboriginal stockmen in return for other desirable items and swapped a drum of molasses (excellent horsefeed in the Dry) for some unusual cane chairs which I still have — though the nails that survived their dip in the briny are nearly rusted out of existence. I should take them to the dump but memories of the heady ecstasy of those long-gone days still linger.

When the Laura Races came around we dressed in our finery. Red and blue velveteen skirts, brown jackets, 'beach material' shirts and dresses. It was a family uniform we wore with affection and appreciation of Lady Luck's beneficence.

Some scavengers weren't as lucky. Treasure was also washed-up on the seafront at Silver Plains and just as eagerly stockpiled and shared but the manager there was more worldy-wise than the naive locals.

"Foot and mouth disease!" he announced and ordered the lot burnt, the little farmyard animals, the snowstorms, the rolls of cloth — the lot.

Nor did the fishermen who flocked to the shoal where the eighty tones of lading were jettisoned. There lay the heavier stuff. Dinner sets, tools, glassware, china ornaments. Customs men were waiting on the Cairns wharf when they unsuspectingly drew alongside.

But for most of the station folk it was an unsolicited bonanza and years later in the Cattle Crunch when prices fell to below the cost of production, when life became extremely drab and colourless and money in more than usually short supply, you often heard a soulful sigh and a mumbled, "Pity there's not another Eastern Argosy".

Marie made her debut as a stud cattle handler with Brafords. We still had three Brahmans, now eligible for the stud — not 'cross' — sections, a bull and two heifers so that we could compete in the Breeder's Group of three 'mixed sexes'.

When we drove to Rockhampton and Cherokee, we also visited Adam Rea's Braford stud and he kindly arranged for us to visit the Tennant and Hinz studs as well. Though Adam is credited with being the originator of the breed, a group of like-minded cattlemen cooperated to put the Brafords on the map, out of the Zebu Cross registry and into history books on their own account. Young sale bulls were scarce on the ground but when we drove back, some of Adam's surplus proven sires were already on the train to Mareeba en route to Crocodile.

What we learnt on our trip to Rockhampton was that now was the time to join the breed societies. So we did. We'd assumed we'd need to build up our numbers first but only a token number of registered stock — and a will to carry on — was required for membership. We'd been members of the Droughmaster Society but Bill loved the whitefaces that reminded him of the old-time Butcher Hill Baldies so we sold some of our registered Droughtmaster females to Bill Edmonds at Wallacevale Stud and concentrated on the Brafords and Brahmans.

Unlike the Brahmans, the Brafords (and Droughtmasters) had to be inspected by a classifier before they were accepted. They were then fire-branded with an official 'C' as proof they'd passed the test. Once dam and sire were classified, progeny could be recorded as calves and inspected at or after two years of age for the next stage of registration. These days calf-recorded progeny of registered parents no longer need to be officially classified.

Getting a classifier to come as far north from Rocky as Crocodile was a little difficult in the early days as the job was held in an honorary capacity by people who had enough to keep them busy at home without foraging off into the wilds. When the time came to take our three hopefuls to the show, they still lacked the all-important 'C'. But we decided to take them, anyhow. There was no class for 'cross' Brafords but the cattle section was eager for competitors and, we hoped, the girls were no disgrace to the breed. Nor was their companion, Graeme, a.k.a Bully, an enormous six months old youngster.

With their snowy heads and necks contrasting with dark glossy coats, they looked rather puritanical, reminiscent of the pictures we'd seen of Quaker women in their stiff white bertha collars. We named the heifers Quaker, Prudence and Constance, the latter, because of her shape and eating habits, better known as Pudden.

Naturally, with no competition, they won all the Braford ribbons and the three Brahmans faced with other contenders picked up an odd one. We did have one heart-stopping moment as one of the ring judges came around the cattle stalls as I was leading Prudence out for exercise.

"That's a nice Braford heifer you have there," he commented, eyeing her off.

I muttered something about her not being a proper Braford as we were still awaiting the classifier but he didn't seem to hear. Mick Costello, the horse judge, was himself one of the first Braford breeders.

We were comfortably settled at the show with an on-site van ordered to arrive before we did. A mare we raced won at Innisfail and, as there was no trophy, Laurie Lewis, the Laura storekeeper, suggested I buy myself one as a permanent reminder. We decided on a cased picnic set that was ideal for our showtime catering. With electric jug and fry-pan we were exceedingly well set-up. A far call from open fires and feedrooms.

The jug was much appreciated by the stock agents and other competitors not staying on the grounds. At the time there was no office for the agents conducting the sales after the judging or for when they doubled as honorary cattle stewards for the show society. The van was usually crowded, the jug permanently on the boil. Laura began school that year as I had back-dated her birthday by eight days so that she conveniently turned five by the first of March. She was keen to start and her true birthday would have meant she had to wait another year to begin. Our Show Day holiday was elastic and took in most of our stay at the show.

Some of the young agents enjoyed playing jokes but we caught them before they had time to fix a boldly printed sign FREE FAIRY FLOSS HERE! to the top of our van. More visitors were something we didn't really need. There was standing-room- only at smoko time as it was.

The main practical joker was Kevin Curry, short, not very slender, with dark hair and dark eyes usually twinkling with mischief. His friend and colleague was Alistair, very tall, slim and fairish, a country boy working for Primac and rather quiet and reserved at that stage. He'd announced his engagement not long before the show.

A group of young bloods including Kevin and Alistair went off after they'd finished for the day to tour sideshow alley. They pulled up outside the strip show where Vanessa the Undresser was drumming-up trade on the platform outside the tent. The showman called for a volunteer but no one offered to get up beside the curvaceous Vanessa in front of the home crowd. The showie repeated his request to no avail. Not wanting to appear unappreciative, Kevin's mates went into action and Kevin was manhandled and hoisted over the crowd to stand beside the scantily-dressed Vanessa. She immediately claimed her prize, seductively caressing him with much gyrating and hip-swinging. With cat-calls of encouragement to both him and Vanessa, Kevin came out of shock and began to enjoy himself. Far from resisting Vanessa's advances he initiated a few of his own with his mates shouting encouragement and advice from a safe distance.

As he and Vanessa were lovingly entwined the showie asked the local boy's name. Kevin withdrew his head from the vicinity of Vanessa's chest and declaimed in that loud, clear voice that has made him such a successful auctioneer, "My name is Alistair Stewart. I work for Primac...."

He got no further. That hands that so blithely threw him upon the catwalk plucked him unceremoniously down and whisked him away through the cheering masses. Alistair quietly detached himself from the group to find his fiancee before his mates did.

The two children loved the show. Lee was a chook fancier and the pavilions with their zillions of entries, chooks of all imaginable shapes and colours entranced him. He came back to camp breathless once to tell of a huge rooster that had been crossed with an emu. It was an Australian Game. Ducks, some squat and apparently legless, others remarkably elongated and tall, pigeons, turkeys, geese and the gamut of 'cage' birds attracted him so strongly we had to be eternally vigilant lest he go there to worship on his own.

Pop Westman always seemed to find succulent, sugary watermelons, even though Cairns Show was in July, one of those few months without an 'r' in them and hence theoretically taboo to melon-eaters.

One year a baby monkey amused us all as it cavorted up, over and through the prime cattle yards. The cattle were apprehensive at first but soon became accustomed to its antics and seemed to watch with as much enjoyment as we did. Its mother was tethered on the far side of the yards near where the showies camped and made frequent irate noises to which Baby was unreceptive. It was fearless and extremely agile, skimming here and there with quicksilvered ease.

Pop cut Lee a tiny slice of melon for the baby monk who accepted it with barely discernible hesitation, quickly scoffed the red flesh and with apparent guilt pangs took the juicy rind to Mum as a peace offering. Mrs. Monk wasn't so easily appeased. Her reaction was typical of many mothers worldwide. Once she had Junior within reach she grabbed him and boxed his ears, took the rind and nibbled it. She then folded him in a monkey hug with her free arm and nuzzled his furry little neck.

Disappointingly for us, it looked as if the little fellow had learnt his lesson. He didn't return but maybe Mrs. Monk was busily entertaining paying customers in another venue. We all missed him.

A mother and daughter team with pretty-as-paint piebald ponies and dogs put their charges through their paces each morning in the stud cattle ring before going to the main ring to entertain the lunchtime crowd. Tom Booth was one of their greatest fans and never missed a practice or a lunchtime performance. He was particularly enamoured of a tiny black and white pony as shiny as enamel and round as an apple. He ending up buying her. For the grandchildren he said.

I wasn't the only one who missed Janice's presence in the days before the show. Others asked after her and were happy she'd got her chance in the Brahman Big Time for Cherokee Stud was right at the top of the tree. Marie and I became more and more confident as we went. The parade went off with no major calamities and the next year we even won a broad ribbon with Prudence against the other breeds. Mr. Atkinson took out the Supreme Champion Cow with his fullblood and Prudence collected the broad green reserve sash. Ken Atkinson also won the Supreme Champion Bull with Wairuna Hilary, Cherokee Festus the Reserve.

Festus's arrival was a bonus. Kalpower Station, up on Princess Charlotte Bay, had bought him earlier in the year with delivery arranged for Cairns Show. His travelling nursemaid was Janice. We soon made room in the caravan and it was like old times.

Bill entered some Daintree cattle in the prime cattle section of the Mossman Show so we nominated our lot as well. Mossman was an ideal family show. It was small enough not to have to worry about the young ones becoming lost, had green grass everywhere, a friendly lot of local competitors, a very high standard in the pavilion and a great place, along a shady creek, to camp. The only thing missing in this perfection was Janice. Mission accomplished, she'd returned to Cherokee.

There were no classes then for led cattle. One for 'Herd Bull under 2 Years" and one for the corresponding female, judged in their pens. To our surprise Bully won the coveted bull sash. Mossman gave beautiful broad ribbons. He was an enormous calf — his mother, Granma, a veritable elephant in Braford clothing — and some of the locals wanted to know was he any relation to Arthur Osbourne's big white-faced bull, so big and outsized that they called him The Freak. He was, Granma was his daughter. We'd brought her from Daintree.

A shadow of sadness hung over the show. A Bush Pilot's plane piloted by everyone's favourite, Bill Fourwell, who doubled as an Aerial Ambulance pilot was missing. Bill was a very experienced pilot and well-versed in getting safely out of sticky predicaments. The plane took off from Cairns but never arrived at its first stop. No May Day call, no sign at all,was seen or heard of it.

Bill Edmonds, down in Mossman from Julatten for the show, said he heard it fly over his place the morning it went missing. It was heading north and the cloud was low — visibility practically nil. His opinion was that it hit the big mountain in front of his house. Other people had heard it flying up the Daintree Valley, a route more to be followed in kinder weather.

Of course, once the plane was lost, the weather began to clear and Sylvia Prior and I, studying Mt. Woondoo across the Daintree River, thought we saw something shining towards the top of the mountain. We rang the Civil Aviation office and they said the parents of a young lad who was flying on the plane to his first job on a Peninsular station were in Cairns. Some of the locals offered to mount a ground search to the sighting and the boy's father wanted to join them. He and his wife had hopes their son and the pilot might be injured, unable to go for help.

It was a false alarm. The unusual shimmering we'd seen turned out to be a huge sheet of thick clear plastic! Weary and disappointed the boy's parents went home to wait. Their wait was long and frustrating. Nothing was found until several years later when a forestry crew going to work a bit later than usual along a track that zig-zagged up Bill Edmond's mountain, saw a gleam in the rainforest below. It was the wreckage. The area had been searched by helicopter and the men had gone that way each working day for those intervening years without suspecting it lay there.

Death had been instantaneous and the bush telegraph hinted there was a repaired generator aboard bound for one of the stations and which could have adversely affected the altimeter reading.

8

Another Good Girl Gone West

Marie was quickly converted from a horse-fancier to a cattle crank. Earlier, while scheming how to increase our purebred herd the children's savings bank balances came into review. They earned 3.25% interest. Chicken feed.

Subsidising them, dollar for dollar, from my Child Endowment and Milkers' Bullocks Fund, we bought two heifers from Mr. de Tournouer. Bill also bought a red bull with a white blazed face and a beautiful, big-framed, red heifer named Bargain. She proved not to be. We found her in her first summer at Crocodile, dead on the other side of the river in the milkers' paddock. Cause of death unknown but she was so fat the ground where she lay was positively oily. She was also close to having her first calf, a prospect we'd dreamed of ever since we'd first seen her. She was in calf to Wetherby Beau Geste one of Mr. de Tournouer's almost legendary sires.

Bill's bull with the white face was bought in an attempt to breed high Brahman-content Brafords. Mr. de Tournouer usually let us know of any whitefaced females he had that he thought might be useful in the Braford herd. Unfortunately most of them were found to have inherited the white on their faces from a 1933 import bull, Supimpa and they usually had calves with no more than Supimpa's facial marking — a wide, white blaze. This bull was different — or so we hoped. And this time our hopes were realised.

His mother was Lady Godiva and she was a direct descendant of a Canadian Poll Hereford Ken Atkinson had imported before his foray into Brahmans and later passed on to Maurice de Turnouer. Canadian Beau Ideal's Hereford markings were persistent. Mated with females with even a mere suggestion of a white face, Lady Godiva's son produced instant Brafords, mostly the required rich red and with diminutive humps thrown in.

The heifers the children bought, Wetherby Coral and Wetherby Brittamart, soon became Sessabi (our Rhodesian friends Terry and Timmy Struckel noted the similarity) and Batty — or the less complimentary, Ratty and Batty. Brittamart, with a huge mountainous hump had little 'bat' ears that showed her Nellore breeding. She also had very recessive genes for courage and for commonsense. The slightest noise or imagined movement would have her up and away in a split second. Sessabi followed her at first but as her commonsense gene was more dominant she soon gave that behaviour away and poor Batty with her recessive courage gene wasn't game to go careering off on her own. Gradually she became less liable to sudden flight probably overcoming centuries of inbred intuition that only the fleet-footed survive. The tigers take the hindmost.

No palatable tree was safe from their dietary habits. Anything they fancied was trimmed as level as an inverted hedge to a height well above their heads. They simply stood on their hindlegs like goats and munched away.

Sessabi was naturally sensible and gradually Batty adapted and overcame her fear of tigers in the undergrowth. We eagerly awaited the birth of their first calves, Batty's a 'full blood', and they duly arrived, born about the same time. A pigeon pair, Cassandra and Bismarck.

The tradition was that the calf's name began with its mother's initial for easy identification. I had secured the services of a retired English maths teacher, Mr. Bell, who'd spent a year supervising the Ahlers children's correspondence lessons at Maitland Downs. We were discussing a suitable name for Sessabi — Coral's second calf, a bull, at the smoko table and Mr. Bell suggested Caractacus. We already had a Brittamart.

"That's right!" cried Lee gleefully. "Cactus!"

So Cactus he was, born with very dominant genes both for courage and commonsense. He was a real pet. As was Cassandra who was number 105 in our herd book. The numbers were growing though most were still 'appendix'. Cactus and Sandra (which fitted in with Sessabi the children pointed out) did their mother — and us — proud. Unfortunately entries for the Breeders and Sires Produce groups needed three and we had only Cassandra and Bismarck. We were confident we'd soon be able to do it in triplicate.

When Marie had been with us a few days an acquaintance asked, "Do you think she'll be all right with that leg?"

I was rather puzzled and decided I'd better have a better look at Marie's legs. She had one and a half extremely shapely ones but one half was very nearly a mere shinbone with foot attached. Marie was a victim of the polio epidemic of the fifties. Knowing Marie, it'd take a lot more than polio to beat her.

Her 'skinny leg' never affected her unless she was overtired and to be on the safe side she refrained from collecting eggs in the haystack if we were away and she on her own. Her 'skinny leg' would snap like a twig if she fell. It hasn't snapped yet — touch wood. Come to think of it, it must've been her left leg, as she was inclined to be a bit lead-footed with her accelerator foot in her youth.

Marie learnt to drive at Crocodile but was already a capable horsewoman and a bit of a bush vet, complete with the family heirloom vet book she brought with her. She was also something of a bush mechanic. After a bit of tinkering we managed to start the veteran motorscythe that I'd acquired from Pop Brodie when I bought his farmlet near Laura.

Janice and I bought a farm each when they were thrown up — during a depression in the tobacco industry, naturally — for tobacco farms. I think they cost us about four or five hundred dollars payable in installments. Pop bought the one next to mine and when he became too old to look after it — and himself — I bought his, too. Among the 'effects' that went with it, were the temperamental scythe and an old cigarette case with a familiar rectangular colourpatch divided diagonally into a green and a white triangle with, inside, a tattered piece of paper with Trooper Brodie's regimental address. The 2nd Light Horse. My father's regiment. By the time I discovered it, Pop was dead. And so was my father.

At times Marie did make mistakes of mechanical judgement like the time she came in spotted like a leopard from the waist up. She'd decided to check the pump engine's oil with the motor running. Or when the stop cable came unstuck on the motorscythe and she cut a swathe through the pineapples before she was able to bring the berserk mower to a halt on the pole of the clothes hoist.

Marie's coming coincided with an exceptionally dry period. Dams were more bog than water and wells dried up. Butcher Hill was sold to the Marsterson family and Bill's parents moved to a small property near Julatten, carefully chosen so that it was on the 'right' side of Rifle Creek, the southernmost boundary for horses eligible for the 'district bred' events at Laura Races. The Boss had his Droughtmaster stud, Kanbrae, named after a successful racemare from his past, and his erstwhile partner in the Droughtmasters, Bill Edmonds, was just down the road a bit at Wallacevale Stud.

The Boss later made a lucky purchase of a broken-down racehorse, a very handsome black called Zarook who, under Dinko Pecotich's training methods recovered and won cups from Cairns to Brisbane. Zarook lived out his final retirement at Kanbrae and is buried there.

At Crocodile we were in a bit of a limbo. Technically we were part of Butcher Hill, apart from the leases we'd taken up to the north and to the west, and were sold with it. Marstersons understood the situation and were agreeable to us trying for a separate lease over what had been our home for so long. It was almost five long and uncertain years before it was granted.

In the meantime Ted Marsterson brought some updated ideas to Butcher Hill. Where he'd come from, Wyandotte on the headwaters of the Burdekin, they were more advanced and had bores instead of the labour-intensive wells of the Peninsula. With the fabulous Mulligan Highway connecting Butcher Hill to Mareeba, anything was now possible. Ted arranged with water-boring contractor Jimmy Johnson to bring his plant up to drill for water. When Ted had enough bores, Jim could put a couple down for us.

Jim owned an old Southern Cross percussion rig that pounded the basalt for hours to gain a centimeter or two of bore depth but, to all of us, it was a truly wondrous machine. Equally miraculous was the operator Jim Cokup who was also a water diviner. And a bit of a showman.

He stomped along, forky stick held horizontally in front of him until the 'pull' of the underground stream dragged the tip of his stick down to point at the ground. Then he began a rhythmic dance going slowly from one foot to the other to measure in some mystic way the depth to the water. He had another incomprehensible rite that greatly impressed the kids to tell if the water were potable or brackish but it was too bizarre for me to recall correctly. I also lack talent with a divining stick.

Naturally the children were all agog and set off with their versions of forky sticks to try their luck. Bill Raymond was a sceptic. His nose and top lip curled ever so slightly at the mention of water-divining but finally he was persuaded to take up Jim's forky stick and walk. He lumbered unwillingly off, fork held stiffly at his waist and a look of disdainful boredom on his face — until the fork abruptly swung down and pointed earthwards. Hurriedly Bill tried to restore it to the horizontal. It wouldn't budge.

Sneaking a look over his shoulder to see if anyone had noticed, he furtively retraced his steps — and the same thing happened. Bill Raymond was converted that day on the road to Spring Creek bore and from then on became a committed water-diviner.

Time was running out for us to get our bores down at a time when they'd do some good. Jim Johnson's plant was drilling holes all over Butcher Hill like an over-industrious grub in a tomato patch. We decided 'to hell with the expense'. We'd buy our own rig. Bill Raymond with his mechanical bent would operate it.

It was a good idea at the time. We put down a lot of bores, mostly with water, put up mills, tanks and troughs and saved our cattle.

On the downside, I lost my offsider, Marie. But that wasn't until later.

We were getting to be old hands at the show game by this. The odd person had even come to me for advice. Definitely odd. As well as going to Cairns Show we'd also nominated again for Mossman. There were new exhibitors coming to Cairns both locals and 'travellers'. Bill Edmonds had his Droughtmasters. Toby Swindley from Mitchellvale at Mt. Molloy had Brahmans. A couple of times we looked after and paraded a Droughtmaster heifer at Cairns for Bill's father. The competition for the District Bred sashes was getting tough.

Bill Raymond lumbered off, having his first go at water-divining.

We took Sandra and Bismarck who were really only big calves — good experience — and a red heifer, Lady Jane, by Bill Edmond's red Wetherby bull, Nucleus. Our Braford team, which easily won the Breeders Group — they were the only Brafords there — was headed by Bully now an exceptionally well-grown fellow not yet two years old. We were able to enter him again in the Bull Under Two Years at Mossman and he won again. Possibly the only bull ever to have won the event twice.

Other newcomers to the Cairns Show were the boys from St. Barnabas' College at Ravenshoe. It was then run by the Anglican Bush Brotherhood and the genial Brother Ted was in charge of both the boys and the bulls. The latter were Brahmans donated to the college as weaners by northern stud breeders. The boys educated and prepared them for show and later for sale with the school profiting financially from the deal. In the process the boys gained experience and had a lot of fun along the way.

This year Dick Fraser had given them a bull, Dundee Celebrate, a well-fleshed bull of whose presentation the boys and brother Ted could well be proud.

Philip Wilson, news reader and news collector for the A.B.C. and the Country Life came around to check-out the cattle exhibits. The boys had Celebrate at the hose, shampooing him ready for judging. Brother Ted supervised.

"Ah," said Phil referring to the scrawl in his notebook. "Ah, he'd be Celebate?"

Brother Ted sank swiftly to his knees, placed his hands in an attitude of prayer, rolled his eyes heavenwards and prayed aloud, "Oh, Lord, I hope not."

In potential sires, such as the bull Celebrate, that little 'r' makes all the difference between a successful future in the cattle game and an ignominious end in a Camp Pie tin.

While we did pick up the occasional broad ribbon sash as well as the more ordinary blue, red and yellow ribbons our main success was with our steers from Daintree. With rainfall and green grass assured there were no setbacks and we won the carcase competition, often taking out second or third with our second pen, six years in a row.

On one occasion our 'worst' pen won with our 'best' pen unplaced having been penalised for being too heavy even though they were well within the age limit. The hook judge commented at the presentation that the upper weight restriction should be removed — or at least not incur penalties — providing everything else was right. He considered the penalised trio ideal. The conditions weren't changed but we learned to take the lighter steers and leave the heaviest at home.

Prior to the advent of the boring plant, the Raymond family sank a couple of wells for us. Everyone — Gordon, Belle, Bill and the girls (Rodney was a bit young) — contributed. Gordon was in charge of the gelignite but it was usually the girls, working in bikinis, who cleared the well, sending the loosened earth up to the top by windlass. One well, despite their efforts, was dry but straddling it with the drilling rig and boring deeper we got water.

One of the first bores we put down was for the house and was divined by both Bills to make doubly sure of a supply. Neither knew much at all about drilling rigs, my Bill infinitely less than the other, so that when the sinker bar swung wildly towards the wooden framework he put his hand out to stop it smashing the timber — and pulped his finger instead.

A very slow trip to Cooktown with Hank Morris, who, hearing of the boring plant, just happened to call in. Bill was back the next day. The tendons were severed but Dr. Pat finally decided against joining them. This left Bill with a straight finger that poked into the boiling water when he went to lift a billy from the fire and which was usually in a state of perpetual damage from getting in the road of just about everything. A lasting reminder of his short career as boring plant operator.

What with one thing and another I didn't suspect a romance was blossoming between Marie and her Bill. Quite the opposite. At times Marie expressed deep sympathy for any girl unfortunate enough to become Bill's future wife.

Her main complaint was Bill's predilection to machinery — the greasier the better — and to a long soak in a hot tub at the end of the day. As either Marie or myself were usually next in the bathroom it fell to us to remove the oily ring from the bath before we showered.

"I pity the poor girl he marries," Marie would say darting a thunderbolt from her dark eyes at Bill should he happen to be within range.

I'd sympathise for I'd done my share as bath cleaner. "Poor thing. I'll give them a black tub for a wedding present."

So it was somewhat of a shock when I'd gone to bed one night leaving Bill and Marie playing old thirty-threes on the record player to wake up to absolute silence — no Jim Reeves, no voices, no nothing — and the dining room light on. I'd turned the lighting plant motor off when I'd gone to bed so the offending light was needlessly draining precious battery power. Reluctantly I got up and went to turn it off.

I got as far as the door.

At the head of the table in front of the silent record player sat Bill and on his knee in an embrace so engrossing that they didn't see me until I gasped in disbelief, was Marie.

"I thought you were playing records," I babbled mindlessly.

"We got side-tracked," Bill informed me in the suavest of tones.

They were married soon after. Marie didn't really want the black tub. She settled for a cheque to the value of.

Things, we believed in the Peninsula, happen in threes. When news limped along the mulga wire that Henry Hanush was gored by a piker bullock (Micky Finn carefully brushed the yard-dust off the exposed kidney fat and stuffed Henry's kidney back in) and Happy Jack was killed yarding-up at Lakefield, we all held our figurative breaths waiting for the third victim. Bill obliged. Helping Len Elmes muster at Olivevale he galloped too close to a tree and had to ride twenty-eight miles (about fifty kilometres) back to the station with a broken leg. This finished that run of bad luck. When Bill returned from Cooktown with his leg in plaster and on crutches we knew we could put disaster behind us and get on with living, hopeful that the next trio of calamities was a long way ahead.

Similarly good things also came in threes. Janice, then Marie and now Alice.

I saw the ad. in the Country Life, 'Girl experienced showing cattle and horses and all facets stationwork seeks position with family'. I got my answer away as promptly as I could hoping no other potential family employer would be quicker off the mark. Alice replied agreeing to take us on and once more I drove down to Daintree to meet my jillaroo.

Alice was from Monto and had more rural experience than either Janice or Marie. She'd been a stud groom with the Clarke's brahmans at Allawah Stud, had a Pony Club Instructor's ticket and knew a bit about farming and farm machinery. As well as that she was a born organiser and soon had us as organised as we'd ever be.

Before I graduated to the more familiar 'Lennie' I was Mother Wallace. Bill was called 'Father Bill', a sobriquet still current.

Hardly had Alice arrived when I left her in charge of my chores and took Nancy down to Rockhampton Grammar to complete her final two years secondary schooling. At that time Johnnie's agricultural college, St. Teresa's at Abergowrie, didn't go past Year Ten and Johnnie pleaded to be allowed to come home or take a job with the Reids at Burnside rather than finish his schooling elsewhere. He came home.

The trip to Rocky also took in a flying visit to Cherokee to see the De Landelles, Janice and Mr. D's unusual bull Ole Red Eye, a purebred Brahman masquerading in a perfect Hereford or Braford coat. Mrs. De Landelles had a soft spot for the whitefaced cattle and hoped to use him in a Braford-breeding program. Red Eye's mother, Virginia, was a white-faced Waverly cow going back to the 1933 import and most of her progeny had at least a white face. Unfortunately the instant-Braford enterprise wasn't as easy as it appeared and Mrs. D's Braford herd didn't eventuate.

Nancy and I drove to Cairns to catch the plane but had we been blessed with Alice's powers of organisation we could have caught a bus at the Y, now Lakeland — the Cooktown-Mareeba-Laura road junction. An enterprising fellow, Bill Wiggins, a.k.a. the Barefoot Busdriver, had begun a more-or-less regular service between Cooktown and Cairns. Bill's bare feet were his trademark and his bus was decorated with a line of barefoot tracks walking up the back of the bus and along the roofline.

Bill travelled along at a rattling rate so much so that Ruth mused 'Billy Wiggins' would be a good name for one of her race-horses. Husband Fred snorted a little at this idea, the road wasn't 'all-weather' and Bill's buses didn't always run according to the time-table. "How'd you know," asked the practical Fred, "when he was 'going'?" Knowledge of vital importance to any racehorse owner.

It didn't take long for Alice to take stock of the establishment. She approved the two show poddies, Ratty and Batty's offspring, Sandra and Bismarck.

"We can do something with them," she announced and we immediately saw swags of broad sashes and satin ribbons appear from visionary clouds.

The three Brafords I'd tentatively selected also passed muster. Graeme, a dark coloured heifer, Hope, and Graeme's young sister, Surprise, whose stable-name was Bubba.

Alice eyed-off with interest the antique baler we'd acquired from the Tableland, into which the hopeful haymaker had to fork hay into the baler's gaping mouth and to tie the resulting compressed bales by hand. Janice made the first bale. It took so long to find the right recipe, we compressed just a fraction too much and as Janice stood proudly for a photograph to commemorate the historic First Bale, one foot resting imperiously on the rectangle of densely compacted grass, the pressure was too much for the baler twine. It — or the knots — gave way and the pose was spectacularly ruined. We were now more experienced haymakers.

"We'll make hay," said Alice.

It wasn't that simple. Before we had a secure hay-patch that show poddies couldn't stray into, it had to be fenced. Herb Wall was fencing on Crocodile out from the homestead and very kindly told us he'd leave the requisite number of coils of barbed wire for our use. We thanked him appreciatively until we found the coils he'd left, just happened to be the ones from the bottom of the stockpile, the ones from which white-ants had eaten the wooden reels the wire was wound on. The loss made it impossible to 'run' the wire off freely from a crowbar spindle. Instead it crocheted itself into a terrible mess of aggressive, spikey wire that spitefully defied disentanglement.

Alice found a solution by cutting it into suitable lengths when her untangling patience ran out. Viewers of our work were impressed by the number of joins to each strain. Herbie was not popular.

We were also to plough — with a pony plough behind the Land Rover until the Lakeland fellows took pity on us and came to the rescue with their State of the Art machinery — a couple of acres for sorghum. The Marstersons sold Butcher Hill to mining man, Clive Foyster, who re-named all but the old homestead and paddocks, Lakeland, after an old prospector who befriended and helped him.

The sorghum had to be in by the end of January so there was no time to lose. When Alice cracked the whip we all responded. Her surplus energy flowed over and we hand-stripped — a rather painful process — bags of urochloa grass seed for which she'd arranged a swap with Justin McCarthy at Lakeland. Grass seed now, for corn and sorghum when it was harvested.

Lakeland was then in its heyday with a Stud Manager, a Stock Manager, General Manager, Agricultural Manager, Machinery Manager and so on, so that the wife of one manager remarked, 'if you're not a manager you're a nobody.'

We frequently visited. The stud, 'Lakefoy', with Graham MacKinnon in charge, was a favourite stopping-off place. One of the conditions to obtaining freehold title for Lakeland was that a Brahman stud be established. Clive Foyster purchased the nucleus of his stud females from veteran breeder Ken Atkinson and in 1969 bought Waverly John de Manso at the Rockhampton sale for a second-top price of $30,000. The sale-topper, Waverly Noel de Manso was knocked down for a then Australian and World record price of $56,000, an astronomical sum for those days.

The stud complex, on the Mareeba side of Lakeland, was most impressive with its white painted rails, scales, and veterinary facilities for Artificial Insemination. Vet. Frank Staunton was in charge of that department and the general health of the stud's inmates. A horse enthusiast, he gave his Arabians plenty of cattlework selecting the cows to be inseminated each morning. There was always something new to see, or some new concept of animal management to consider when we visited Lakefoy and we were assured of a cup of tea at Bev MacKinnon's before we left, inspired and up on cloud nine, for home.

Alice was a persuasive and unashamed borrower, and appropriated both ideas and material goods. She also latched-on to the radical conception of Managerial Conferences — a mysterious event that happened very regularly at Lakeland, the biggest privately-owned grain farm in the southern hemisphere.

Managerial Conferences became our excuse to stop what we were doing — boring work — and to have a cuppa and a chat. It made Alice's taskmastering a lot easier to bear and we got lots done.

I'd be welding gates for the feed pens or making hayracks with my 32 Volt welder, with Alice painting the finished articles and commenting rather scathingly on the amount of 'seagull' welds (meaning they were lumpy — like bird droppings), when she'd down tools and announce, "We'd better have a Managerial Conference. "One of us would return to the house to make tea while the other finished off the job in hand.

Fortunately for me, Johnnie was conned by Alice into helping hammermill the sorghum when it arrived from Lakeland. It had been treated with an insecticide, Phostoxin, that brought both Alice and John out in an irritating rash that persisted for weeks, despite using all the allergy pills in the Aerial Ambulance kit plus my entire stock of calamine lotion and metho. It did, however, keep the weevils at bay.

The same partnership went into the manufacture of supplementary blocks to be fed out to the cattle in the Dry, made to Alice's recipe of salt, molasses, a smidgin of hammermilled grain and various minerals. No container, be it cardboard or metal, was safe from their predatory paws. Twenty litre drums cut in half were ideal and once they were emptied could be re-filled but in the meantime any good-sized containers I'd stashed away for pot plants or trees had to be guarded zealously by day and by night.

Alice was a master at presentation both personal and bovine. Shortcomings were quickly recognised and steps taken to camouflage them so successful that on judging day Alice's usually scruffy work-a-day self and the normally rough-and-ready show poddies, all scored close to a Perfect Ten. Standing cattle correctly to maximise their strong points and to minimise or totally cloak their bad ones, was an art Alice practised consummately.

The poddies were subjected to leading, 'standing' and ribbon practice, though at first some of the more naive still attempted to eat the ribbons, and all had to have rugs, 'pyjamas' as Alice called them, to keep them from growing long, unattractive winter-coats. There were dust-sheets for daywear and thick, heavy, lined bag rugs for night attire. All the latter I made by hand with speying twine and a bag needle from corn sacks, though Alice effected any running repairs.

Graeme was now a very big bull. It took two corn sacks to get the correct length for his rug and though he'd grown out of his last year's one I was rather loathe to put all those stitches into such a bulky new one. It was heavy work sewing pyjamas for him and he didn't have a second thought about rubbing against rug-destroying objects should the mood take him.

Alice argued her case but I resisted. "When he weighs a tonne," I compromised and again there was dissension over that. I still have the memo she placed under my breakfast plate which stated ' Bully's measurements (no dinner). Waist nine foot three inches, girth eight foot two. Head top to tail butt ten foot flat'.

Alice argued that he did weigh his tonne. Our scales which worked on a fluid displacement system only weighed to that amount and of late, when Bully stepped on the weighing platform, the result was extremely difficult to read and the marker returned to anything but zero when he got himself off. Still, a tonne was such a very good weight for a bull with the bare minimum of show preparation and under two and a half years old that I decided Alice's argument was just wishful thinking. I resisted successfully for some time but finally had to give in and make him his new rug. Alice seemed to be strategically remiss at mending the tears in his old one and soon there were more holes than sack bag to keep Bully's coat in order.

Another innovation Alice introduced was 'Sunday Off'. After Sunday breakfast there were no lessons for the show poddies, their pyjamas were removed, day sheets left folded on the rail and once they realised what was 'on' they bucked and frisked down the paddock for a whole day's freedom. We didn't have to get them back in the evening. They were always waiting impatiently at the gate for their tea. Freedom is a wonderful thing, it's true, but tuckertime has an equally strong appeal.

Once again we tried to get a classifier up from the Braford Society in Rockhampton more especially to give our show Brafords some authenticity with a Society brand of approval stamped on their hides. The delay was understandable. We were an awful long way from Rocky and the classifiers did their work in an honorary capacity at that stage, during rare lulls in their own daylight-to-dark work schedules.

This time, we had a date. Bill Rea would fly up to Cairns on March 26th at 2pm. I was to meet him and bring him first to Daintree and then on to Crocodile. Our show poddies — should they pass Bill's critical inspection — would at last be legitimate Brafords. When the telegram arrived, the skies at Crocodile were blue with no rain in sight, but, true to our classification form, Jim Boileau, expected up with a buyer for whom we'd been mustering cattle for the last week, phoned from Mareeba to say he couldn't come. The Mary was over the bridge.

If he couldn't get up, I couldn't get down. A hasty telegram was sent to the Braford Society and Bill Rea's destination was changed to the airstrip at Laura. The arrival time was the same and less distance for me to travel. Great.

Bill turned the would-be sale cattle out into a holding paddock and got the Braford mob in ready to be drafted-out first thing next day. Next morning we kept the show poddies in close-handy and I dithered around waiting for the call to go and pick up Bill Rea.

The call I got was another telegram. 'Due bad weather Rockhampton trip reluctantly cancelled. Bill Rea.' He hoped to make the trip up before Cairns Show.

We hoped so, too. In the meantime Alice kept everyone working and the poddies heavily involved in higher education. I made rugs and more rugs for cattle and some jeans and shirts for myself — so I wouldn't let the side down.

All was going well until a letter came from the Hostel in Herberton saying Lee's school work was deteriorating and they suggested I bring him home. The idea of sending him to Herberton the previous year was so that he would have company instead of being the only child at home. This had worked well while Nancy and her friend Frances Gostelow were there, as they acted as de facto mothers to the few littlies. With both of them gone Lee and his contemporaries felt deserted and most forlorn.

I went straight down and rescued him and he did look neglected and waiflike. Luckily, that was only transitory and he was soon back to his usual self — with some noticeable improvements. Teaching him was much easier. He'd been broken-in to the routine of class work and been weaned off some of Mr. Bell's eccentricities. He handled the correspondence lessons with such ease that he often had afternoons free to offside for whoever needed an offsider. It was a move for the best and good to have him home again.

For some time Percy Trezise had been reconnoitring around looking for aboriginal paintings and artifacts. Percy usually spotted likely places for caves from the air in his role of airline pilot and later came in on foot to look closer. The only problem was that Percy was usually so short of time for these forays that he didn't let us know where he was going, so that if we came across the remains of a campfire or some man-made sign we dismissed it as evidence that Percy, like the mysterious Foo, had been there.

About the same time a few choice Brahman-cross heifers began to turn up missing so we came to doubt if all the tracks were Percy's — or if they indicated a possible poddy-dodger. So, when on a round taking salt blocks out to the troughs I struck Percy at Crocodile Hole I told him (rather heatedly) of the dilemma. Percy saw the point and agreed to 'give notice' when he was on Crocodile. He then introduced me to his companion. It was Xavier Herbert. Xavier was interested in finding William Hann's camp which he thought must've been somewhere near Crocodile Hole. Hann made his epic trip from Georgetown to the Stewart River which flowed into Princess Charlotte Bay, crossing the Palmer/ Laura River divide. This made the permanent waterhole at Crocodile a logical choice. Hann named the Laura River the Hearn, after his wife's family, but missed out to the second 'discoverer' who named it Laura for his wife. Hann returned, safely though near exhausted, via the heads of the Normanby River in the early 1870s.

One of the aboriginal paintings Percy had re-discovered was of a gigantic red ochre horse and they decided it could only have been painted after seeing Hann's horses pass. The Jardine brothers, like the tragic Kennedy, went much further westerly on their way up the Peninsula.

Percy's next stop was the Byerstown Range where he was trying to find the site of a meteorite strike some years before I came to Butcher Hill. Bill had a small piece of meteorite his mother used as a paperweight but after I lent it to an interested geologist who was going to confirm its meteoritic origins, we never saw it again. I can only gather from that, that it was indeed a dinky-di meteorite.

Later Percy called on another trip and left some of his gear in one of the sheds at Crocodile to save carting it back and forth to Cairns and he also took me up to show me the caves he'd found, in the hopes that I'd be a sort of honorary ranger, sorting out the sheep and the goats when people asked directions to the caves. Only special people were to be shown Quinkan, Giant Horse, Ibis etc galleries. The hoi polloi were to be directed to the easily accessible Split Rock discovered early on the side of the road by the Main Roads men.

Percy is a marvellous storyteller and when he told of his first trip to Quinkan Cave with his mate Dick Roughsey on the day — unknown to them beforehand — of an eclipse of the sun, he made the hair on the back of my neck stand up just listening to him. They thought the quinkans had got them when the daylight turned to dark and, even in broad sunlight, I wasn't that sure I was safe from them either. There was an eerie feeling about the place. You could sense the presence of other beings at all times. I could imagine what it would've been like in total darkness.

With my interest greatly enhanced by Percy's stories, when May holidays came we took the children to Split Rock and also to a place Sam Elliott had shown us on the 'old' Laura Road, near the coal seam, the Chinaman Tree.

This tree, a Leichardt with limbs radiating almost horizontally from the trunk was a Chinamen's night refuge. Iron spikes provided footholds of sorts from the ground up the trunk and to the lower branches. Here, Sam said, they had made platforms on which to camp, safe, it was hoped, from flying spears. Stones were piled for convenience (and protection) on the platforms but very little was left of these and the intended stone weapons were probably the ones gracing the ground underneath.

Since Butcher Hill had been sold — and us with it — Bill was unsettled at Crocodile. The time it took to have it took for us to have title for it included in the two leases we had on either side of it dragged on and on. Just when we thought things were going right, some other obtacle would be discovered. We took up the block with the Quinkan galleries as vacant Crown Land and bought the other portion, taking in part of Kennedy Creek and the Mossman, from Sonny MacDowell when he wanted to sell. For some time we had used it in conjunction, splitting the rates and rents, us using the Kennedy side and Sonny running his cattle up the Mossman.

We heard on the grapevine that Southedge, near Mareeba, was for sale. It was on the drovers' road to the Mareeba saleyards so Bill was fairly familiar with it. It was probably better country than Crocodile with the added advantage of being closer-in and a shorter distance from our Daintree block. The price with five hundred head of cattle was three hundred thousand pounds, a snap these days but then a considerable amount of money.

Reluctantly we decided we'd have to try to sell Crocodile first to raise the finance. The best of the cattle we would take with us.

Picking the children up from Herberton in a there-and-back effort in the one day, and then having to drive to Cairns to pick Nancy up from her train, I looked longingly at the Southedge turn-off each time I drove past. Wouldn't it be lovely! Practically in the suburbs. No early starts and late finishes and busted tyres, electric pumps that failed to pump, melted coils and/or condensers and the run of things that usually went wrong on these trips. Thinking about it, the road home from the turn-off seemed longer, lonelier and bumpier.

The group who were interested in Crocodile came up to inspect. They flew from Victoria to Laura in an amazing seven hours. We had our fingers crossed. The agent handling the Southedge sale told us the vendors would accept a cash offer of two hundred and fifty thousand dollars. The suspense was unbearable but we couldn't come up with the cash. Our interested buyers decided against Crocodile but bought instead a line of bulls for Arukun, then a mission station with a large cattle herd.

We didn't have time to be disappointed. A phone call told us Ned Cobb's place, on the 'right' side of the river to Louis's old place at Daintree was for sale and we had the first offer. That was a decision we could make quickly. Ned's place, Glengarry, would complement Louis's Woondoo perfectly.

Meanwhile, a little distant from the hectic scene of buying and selling — and dreaming — Alice had the show team blooming. She decided they needed just a suspicion of a hair trim. A few coarse hairs on the tops of their heads and in their ears detracted from the overall sleek and shiny look. Typically, she'd let it be known she needed horse-clippers to do the job, and, just as typically, they materialised. An old horse-trainer friend, Toby Kelly, had a set he no longer used. We could have them. He'd give them to Sid Elmes to bring up.

A welcoming committee was on hand to greet Sid who unloaded, with the air of a magician pulling the white rabbit from the hat, a tall metal thing with a handle and a length of cable. Used to the conventional 32V human clippers we novices were puzzled. But not Alice.

She pounced on the strange contraption with cries of rapture. It was clippers — but not the electric kind. These were activated by one human (me, mostly) tirelessly spinning a handle at the required numbers of revs, changing arms smoothly but regularly as the arm engaged in the action threatened to become detached from the aching shoulder — while another person (Alice) delicately manoeuvred the cutting head over the bewildered animals' polls and even — if she was lucky enough for them to stand and permit the intrusion — in their ears.

Alice knew all about this ingenious invention and soon had it running much more smoothly than they did in her initial attempt on an uncooperative Bully. He protested politely but successfully by raising his head just beyond Alice's reach. Once done, the effect was very pleasing. The poddies looked extremely chic and so much more aristocratic and highly-bred.

We had two more 'firsts' that year too. Civilisation certainly reached out and grabbed us. One vehicle brought an evangelistic team of Jehovah's Witnesses and a later one with Government registration plates disgorged two agricultural economists from the then Bureau of Agricultural Economics, now A.B.A.R.E.

George and John were doing a beef survey and asked a host of questions for which we had to find reasonably credible answers. At times, even Alice was temporarily stumped but she rallied and came out with imaginative but apparently satisfactory figures.

At what age did we mate heifers? What percentage of bulls did we use? At what weight did we wean? At what age/weight did we turn off steers and cull cows? Percentage of weaners to branders?

Our husbandry practices were governed in the main by bank managers and seasons but the economists needed numbers and percentages to draw graphs to show what we'd done.

Crocodile had sufficient paddocks to allow for controlled mating, so that calves were not born in the annual Dry season — a nutritional drought that came yearly — nor to very young heifers who still had their own growing to do before motherhood overtook them. We hoped our confident-sounding guesstimates would satisfy the economists' needs but were more than mildly surprised when they said the next stop was Colin Gostelow's Violetvale.

Violetvale, much farther up near Princess Charlotte Bay was a long, long way from fencing supply shops and had few fences. The bulls' individual libidos controlled the matings and the main criteria for sale cattle was any you could muster and get to Mareeba in time for the sale. Paperwork and percentages weren't high on Colin's list of priorities. We wondered how they'd found Violetvale existed and were sure their visit would open up interesting possibilities.

In due course George and John returned. The outward change in them was easily discernible. Gone were the neatly pressed trouser creases and finely starched shirts. Gone were the dark blue ties and the mirror finish of their laced shoes.

In their place, rather crumpled khakis that bore little resemblance to their former spotless state. A closer look showed blackened smudges on trouser leg and sleeve. There was an unsubtle personality change, too. Self-assurance seemed to have slipped somewhat.

"How did the questionnaire go?" we asked over a cup of tea.

The pair exchanged looks before one replied in a careful but casual, off-hand way, "Oh, we got side-tracked."

Colin, having very little idea of vital bovine statistics and no inclination to pursue the avenue towards enlightenment, took them pig-shooting — something he knew a bit about. Wild pigs, wild bulls and wild horses were more in his province. (Some might add wild women but that would be a libel.)

Unfortunately a fire had gone through the tea-tree at the edge of the marine plain just before they went on their safari, blackening the paperbark of the slim saplings with a friendly soot. Taking to the trees with a wild and wounded boar in pursuit, George and John worried not about a little charcoal blacking their clothes and their persons. The main thing was to escape the huge tusks and the obvious evil intent of the boar that treed them.

Colin, of course, found the situation hilarious. In his opinion it was only a very little boar, not much over sucker size. Certainly not dangerous. As the fugitives attempted to climb up higher from the irate pig's range, the sapling bowed disobligingly downwards until they hovered unsteadily, barely clear of the gnashing jaws.

When he'd had enough entertainment for the afternoon, Colin shot the boar allowing his visitors to descend in safety and produced his rum bottle to restore their frayed nerves.

"It was incredible! No one will believe us when we get home!'"

"There are some things," John said with great dignity, "that are perhaps better left untold."

For all that, we thought they enjoyed their experience — once they were rescued — and the opportunity to see how the legendary other-half lived.

In due course we opened the mailbag to find a small red (ox-blood red) soft-covered booklet full of mystifying graphs, figures and percentages and — for all we knew — hieroglyphics, purporting to pertain to the marketing of beef cattle in the Cape York Peninsula and signed "With the compliments of the authors." My first author-inscribed book.

Back to the routine. Alice decided we should have a 'sign', a sort of logo, something we could take around the shows with us for advertisement and identification or leave on a tree at the roadside mailbox. I was given the afternoon off to design it and, surprisingly, the design was readily approved and soon transferred to a spare piece of flat galvanised iron — part of the G.I. containers roofing iron then came in.

There was a simple shield, broad at the top, pointed at the base, divided into roughly thirds with a Braford bull's head and a Brahman bull's head facing each other in the top two-thirds while a greeny-brown crocodile had the remaining space all to himself, sprawling menacingly across the bottom. Once it was painted we rushed out to hang it at the mailbox but in the hustle of departure at showtime no one gave it a thought and we went the rounds without our identifying logo. Alice slipped-up there. It lasted quite a while as those were still the days before travellers were infected with the desire to shoot up any available roadside target — animal, vegetable or mineral.

9

Showing in the Rain

The importance of Laura Races as our social event of the year diminished somewhat when there were show poddies to cosset. We couldn't take the time off to camp at Laura as we used to do but with the road improving every year it didn't take long to drive down and back each day. As a sign of how the years roll on, one of the erstwhile 'little Cobb kids', Val, was Belle of the Ball and well deserved her win for her appearance and her personality. Another old friend paid the races a visit, too. Shirley Porter, once of Olivevale and now Mrs. Ken Robertson with the exotic address of 'Phoenix, Arizona'. It was great to see her again and to reminisce of the 'old' days when Shirley had the reputation of being an excellent but wild-and-woolly horsewoman, preferring to jump the sliprails on her trips to and from Laura township rather than use the more mundane method of egress — letting down the rails.

We were to take a Droughtmaster heifer of Bill's father's with us to Malanda show but she became temporarily lame after an argument with a barbwire fence and had to be left behind. I'd sewn a rug for her but that was no waste of time. We could always use spares.

George Kelly from Springhill, one of Toby's two sons, agreed to take the cattle for us in his truck. This left the ute free for our gear and as many bales of hay as could be imaginatively loaded on and tied down. Alice went with George and the cattle. Lee came with me.

Typically, as we got within a few miles of Malanda it began to rain, non-stop, proof that we were heading in the right direction. Rain for the show is a Malanda tradition. The stalls, on the other hand, were roomy and bone dry, so the cattle settled comfortably in their pyjamas to have hay and grain in bed.

One of the stewards, Percy Kidd, an old hand and very knowledgeable, suggested we book into a cabin at the caravan park next door. Good thinking. He showed us a short cut we could use when checking the cattle. Like all short cuts it had its drawback. We had to cross a huge drain, reminiscent of pictures we'd seen of the Grand Canyon. We must remember, Alice advised after we clambered across it the first evening, not to come home drunk.

The rain didn't let up for a moment. It came down in a grey blanket that tinkled rather melodiously on the tin roof of the cabin and lulled us to sleep in no time at all.

And woke us up again at daylight, a little more peremptorily, to remind us it was time to take our charges their breakfast in bed. It was freezing for us tropical birds and we wore as many layers of clothing as we could get away with, plus our ubiquitous gumboots.

The cattle didn't look as happy as they had the night before. They weren't drinking. They thought they'd absorbed sufficient moisture from the atmosphere. All were tucked-up and hollow and shot us nasty looks suggesting that more than one layer of clothing should be the rule for them, too.

Lee, tired of the inaction and the weather, leapt at the chance to desert us for the company of his Merluna cousins, down with their grandparents for the holidays.

To relieve the monotony of the rain, someone came over to ask would we take Bully and Bismarck, Brittomart's son, over to the bar. Jeff Bates (Mr. Dame Zara) was there and expressed an interest — but not great enough to incur muddying the boots. With some trepidation we sloshed them over in the mud and they behaved beautifully, even if Bully did curl his lip distastefully when offered a comradely stubby.

Alice, eager to prove her point, brought some of the dairymen in to back her opinion that Bully weighed over the tonne. Dairy bulls — A.I.S. and Friesian — are big-framed animals and when they pile on the condition squiring their wives about in lush always-green pastures, often 'do their tonne'.

They assessed Bully's length, (two cornsacks, as I've mentioned) his depth and his breadth, did some mental geometry and came up with the unanimous decision he'd 'do' his tonne easily. A small victory for Alice but I was still not firmly convinced.

We won the classes we had them in, and the championships. Unlike the dairy section where competition was extremely fierce, there were few entries in the beef cattle judging in those early years.At Atherton we were joined by Bill and Linda Edmonds with their

Droughtmasters. It was still raining but there were odd moments — very fleeting — when it didn't.

United with Lee again, we made a comfortable camp in the cattle stalls taking an adjacent one for a feed room and screening the both for privacy with tarps. We heard some interesting conversations secluded away behind that tarpaulin curtain.

Alice suffered extreme physical distress controlling the urge to giggle when a group of primary schoolboys pulled up in front of Bully . We overheard a lot of admiring ohs and ahs and a considerable amount of whispered asides until one strident, high-pitched voice rose above the hum, "Gawd, look at the size of 'im! 'e wouldn't 'alf flatten the 'eifers, would 'e!"

The one negative part of our stay was the shower arrangements. The only showers were the ones used by hardy footballers in the playing season and the only water available felt like it came straight from the ice-works. Linda, Alice, Lee and I would troop up, one stand on guard at each of the two access doors while the other two ducked swiftly under the ice-water, lathered, rinsed, dried and dashed out again. It seemed so much warmer for quite a long time after one of those brief but bracing showers.

We shared the top cattle honours with Bill and Linda. They took out the Supreme Champion Female with Calamity.(I apologise for not recalling her real name but she was rather accident-prone and her nick-name was appropriate.)Bully took the broad ribbon in the bull section and Bubba took the nail almost off Linda's big toe when she put her hoof on Linda's boot and then abruptly changed direction.

We were too busy during judging — and applying First Aid to Linda's toe — to notice the rain was easing. Perhaps in our raincoats, Akubras with downturned brims and gumboots, we'd become immune like true Tablelanders, but the sun came out in a spectacular fashion for the Grand Parade. The dairy cows and their white-coated leaders almost dazzled and our mud-bedraggled crew were once again spotless and gleaming. It was an impressive display, a long snaking line of cattle, horses and riders parading in brilliant sunshine on the brightest of green grass.We didn't have to parade all ribbon-winners which meant we needn't seek out polite but reluctant volunteers from our acquaintance who most likely lacked experience or who had no desire to gain it in such a public spectacular.

Alice took Bully. I led our Senior Cow, Jane. Bill Edmonds took his champ while everybody's friend, Jim Boileau, paraded Bill's bull. Poor Linda watched from the sidelines her foot impressively bound up in an infinity of snowy white bandage.

* * *

George arrived on time with his truck and with more hay. The cattle and Alice left with George via the winding Palmerston Highway . The gear, mountain of hay securely roped down and canvas-covered, followed with Lee and I in the ute. I add here my passenger was not at all happy. He did not see why Alice should be allowed to go in the truck every time. In case of fairness, it should have been his turn. Not speaking to me, the author of all this injustice, he retreated to his corner of the long, bench seat and muttered. From time to time I caught snatches like, "I'm not going home with you, anyway. I'm going with Dad. In the truck."

With a mother's pseudo-confidence in her power over her offspring I took no notice. He'd get over it.

We arrived at Innisfail to another refrain of rain on the roofs, mud underfoot and the pleasure of seeing our new quarters. Large roomy horse-boxes. Two rows with a wide aisle between. Innisfail show was held at the racetrack. We were allotted the luxurious horse-boxes, unloaded, led the cattle to their new quarters and made ourselves most comfortably at home in a stable across the aisle from them.

Lee and Alice had narrow wood and canvas folding stretchers. I unrolled my swag on a bed of hay bales and probably had the more comfortable bed until our stay at Innisfail became prolonged and my bed grew shorter each time the animals were fed. George left, to return to take us to Cairns when the judging was over. Lee went exploring. I re-organised the human and bovine catering services. The sun came out and Alice gave the show poddies the bath of their short lives.

It had been getting increasingly difficult to tell Bully and his mates were white-faced Brafords and not all-over-red Droughtmasters. Armed with lots of show shampoo, grim determination and the beautifully soft Innisfail water, Alice set about to remedy the situation. She'd finished the greys and started on Bully when I joined her. For once her air of superb confidence was shaky.

"Look," she said pointing a trembling soap-lathered finger at Bully's face.

It was so dazzlingly, radiantly sheeny-white, I couldn't see why she was acting so strangely.

"Look," she repeated, jabbing at Bully's forehead.

I saw. Horror of horrors. His face was not pure white but mottled! The washed hair, so squeaky clean as to be transparent, revealed small red polka dots on his gleaming skin. As if we weren't damned enough with an un-inspected Braford at a show, we now had one with prohibited spots.

Our first impulse was flight. Leave the critters. Get the hell out of here and away from them. Our second was more rational. Review the situation. The verboten spots that we'd previously had no idea even existed weren't in the hair but on the skin itself. Most unusual. Despite efforts to think up a solution, we were stunned.

Alice recovered the quickest, thrust Bully's lead into my trembling hand and sprinted for the camp.

Almost immediately she returned with a tin of baby powder — another trick she'd learned in her apprenticeship — rubbed the talc into the offending polka dots and towelled the hair dry over them. We no longer had spots before the eyes, but Bully's face did look a little like a freshly Blanco-ed sandshoe.

Horror of horrors! A Braford bull with prohibited spots.

We finished the others, taking care not to overdo the face scrubbing. No more damned spots. Then down came the rain. Time, Alice suggested, for a managerial conference.

I agreed.

While we were inside, warm and dry, sipping our tea and discussing the next move, a truckload of dairy cows unloaded. Friesians, black and very red, rather than the conventional black and white. We watched as the woman in charge of them, like us a first-comer to Innisfail, unloaded in the rain. Unlike us, in her raincoat, rainhat and gumboots she gave the appearance of being well used to wet weather. She tied her charges to a handy rail and delved into the truck's cab for her cow-washing gear.

No sooner had she led one over to the hose and concrete slab we'd just vacated, when an enraged female flew out from the horse stalls and verbally attacked the would-be cow-washer. We knew her and her reputation and spinelessly made ourselves as silent and as invisible as possible in the hope she wouldn't associate us with any of the remaining soapsuds. She must've been somewhere well away when we did our washing for she was a very vocal woman.

The hosing block, the Amazon informed the dairymaid, was for horses, not for cattle, and with much blustering and arm-waving soon had the Tablelander and her cows in retreat.

Like the gutless wonders we were, we held our breaths and said nothing. The cows faded a little over time in the drizzle but never completely lost their red tint. The Friesian owner vowed she'd never come back to Innisfail, but as the Amazon also travelled the northern show circuit we decided just striking Innisfail from the list wouldn't solve the problem.

Lee mated up with a small boy his age, a budding entrepreneur, and they were off to a successful partnership. Lee was initiated into the monetary advantages of collecting refundable softdrink bottles. In no time the area around the stalls and the machinery display was bottlefree. Any non-returnable duds were disposed of in the bin before they raised their hopes a second time and in rare sunny breaks the boys could be seen sitting on a favourite miniature garden tractor in a machinery display counting their money. I think they may have had hopes of buying the tractor but bottles weren't quite that prolific or profitable.

The judging went well for us. It was held inside the spacious stable complex — far too wet outside.The judge, an ex-D.P.I. man who married a local girl, now bred his own Brafords down Nebo way. His judging was easy to follow and consistent. If you were beaten you knew the reasons why and if you won, your animal's finer points were clearly enunciated. Anyone could learn from his comments.

Bully ended up as Supreme Champion beating our old friend and Show President, Fred Drew's Cherokee bull. The spots were forgotten. In fact, with a bit of dust and day-to-day accumulation of dirt on the facial hair, they'd disappeared (though we now knew they were there and lurked in waiting.)

Despite the beautiful soft water, the cattle still wouldn't drink — even with molasses added — and were looking decidedly tucked-up. The Friesian lady was having the same problem. Her cows' sides were almost touching. Our horseperson had frightened them off anything remotely wet.

The Grand Parade was a leisurely stroll of whatever cattle and horses owners wished to parade down the straight of the racetrack. The beef cattle led the parade, dairy cows next and then the horses, with a strong thoroughbred contingent, at the rear. Again the sun shone. I took the lead with Bully and his broad tri-colour sash and Alice followed with the white 'twins' one in each hand. She'd taken advantage of the brief sunlight and taken photos of the cattle with their ribbons.

I was heading the parade, not overconfidently but well enough, approaching a long pool of rainwater at the end of the straight when a rattle of hoofs came from behind.The Friesians had spotted the water and raced ahead slurping it up while the supply lasted. Bully elected to join them rather than pass up the treat but Alice led her two past and took the lead. Sated, Bully and the Friesians, now rounded and happy, contentedly re-joined the parade at the end of the cattle section and we returned circumspectly to the stalls.

Temporarily. Lee and his partner were there counting their takings. Alice and I were invited to the official show lunch but we'd brought back something for Lee and his mate to eat so we left them to it, quickly spruced ourselves up and left for the luncheon.

John Boydell, master photographer and later Droughtmaster Society stalwart, sat with us and shamelessly encouraged Alice to filch salt and sugar in paper serviettes for our reducing larder. He even hailed a passing waitress for 'seconds' which he donated to Alice for our tea.

Our other old friend, Fred Drew (whose bull we had beaten), as Show President kindly welcomed us in his luncheon speech. Later he asked if I'd consider taking on the cattle judging the next year. I was flabbergasted. Alice assured me I couldn't be much worse than some she could mention but I declined the honour.

Luncheon was over and with it the sunny break ended. We made a dash through the rain for the stalls only to be puzzled by a large crowd filling the laneway near Janey's stall. The reason was soon evident. Lee had improvised a bridle from a halter and two lead ropes and was riding Jane with his mate double-banked behind him chanting, "Penny a ride. Penny a ride.Who wants a ride?"

I don't think they had any financial takers. They slid off quickly when they saw Alice and me and tried to cultivate an innocent, aloof look. The onlookers must have enjoyed the diversion as it took a while for the crowd to disperse.

The show ended but not the rain. The Friesians went home immediately after the parade. Some of the horses took the road out to Cairns. One truck getting bogged at the ramp had to be towed out. The front of the loading ramp was such a quagmire as a result that George decided to take a chance and wait until morning.

The new day brought more rain. And more after that. My bed shrank to hip-length as the hay bales disappeared down the animals' necks. Alice, George and Lee visited the corner store for more tucker. I went to my cousin Brenda's place to wash and spin practically dry our show ring clothes in the hope they'd be ready for Cairns, wondering over and over would we ever make it out of Innisfail to get that far.

Would we be able to get our favourite stalls at Cairns? We were doubtful. The rules were first in, best dressed and we well knew it. The continuing rain increased our concern.

In the meantime our judge was also weather-bound and decided to have a second look at the remaining cattle to help pass the time. Judging over, it was now permissible to fraternise so we invited him for a cuppa.Tactfully we dodged the subject of Brafords in general and Bully in particular and were struck dumb in our patter of inconsequential small talk when the ex-judge remarked casually, "A good bull is always a good bull even if he does have a few freckles on his face."

I wasn't game to look at Alice.

So much for the baby powder ploy.

* * *

On Monday it stopped raining in Innisfail and we made it to Cairns where it was bright and sunny. Kalpower and Bill Edmonds were there, also our caravan, and we were able to get stalls that suited us. On top of that, the astronauts landed on the moon. We saw them on a T.V. set up for public viewing on the showgrounds. All this gave credence to the old adage that miracles never cease.

We had a day to get organised before judging started. The cattle found the Cairns water more to their liking but when we weighed them they were well down on their weights before they left home. Even so, Bully weighed 2020 pounds and I had to apologise humbly to Alice (who had never harboured the slightest doubt that she was right).

I'd been making rugs for the children from the broad felt sashes and needed some more greens (Reserve Champion) and the tri-colored red, white and blues (Supreme Champion). Naturally, with no competition, we got the Braford sashes and no one seemed to notice the cattle didn't sport the inspector's symbol brand. One of the 'twins', Cassandra, was Junior Champion (a purple sash for Billy Kid's rug), Jane was Reserve Senior Champ and Bully's sister Bubba was reserve to the Supreme Champion Female - Bill Edmonds' Droughtmaster, Calamity. But there was to be another green sash for the rug. The judge called both Bully and Mr. Atkinson's bull Indus out for the Supreme Champion. Alice led him and had him standing beautifully but I was a bundle of nerves. On one side of my brain I could see the headline, BRAFORD TAKES OUT SUPREME CHAMP CAIRNS SHOW. On the other I could see BREEDERS EXPELLED FOR SHOWING UNINSPECTED STOCK.

The judge walked a dozen or more circles around the two bulls, stood back, thought, changed direction and circled again. Finally he indicated to the steward Indus was to get the champion sash. A sigh of relief and we had another green, enough with the tricolour we got for tying for first with the Osborne brothers in the carcase competition with Bill's steers to finish Nancy's rug. It was the last time we put steers in the carcase competition but the tie was enough to count as a third win and give us our second cup — outright.

The Grand Parade went off without a hitch. Lee still insisted he was going home in the truck and the last I saw of him as I left hurriedly with Johnnie for the airport to pick up Mr. Rea, he was sitting in the truck waiting. Bill was to load up early with cattle for Daintree and then bring the show cattle up to Bill and Kit Edmonds' place at Julatten. We were to overnight there.

Bill was at Edmonds when I arrived with Mr. Rea and Johnnie. But where was Lee?

"I thought he went with you," said Bill.

Oh, no! Julatten phone was a manual exchange that did not work at weekends but there was a phonebox outside the Post Office which did. Kitty and I drove up in record-breaking time and told the poor harried show secretary my problem. No lost children had been delivered to the office and though it was now quite dark, he'd send someone down to the cattle section to have a look. We were to ring back in an hour.

What a long hour! We phoned back on the dot to be greeted by a familiar voice (Bill and Kit's niece, Dianne's) saying Lee was found and they'd bring him home in the morning.

Lee was in the truck when I left but soon afterwards his cousin Harding came along and off they went for a last look at the sideshows. Of course, when Lee returned there was no truck but, parked in the cattle section near the caravans was a familar-looking Land Rover. He climbed in — no one locked anything in those days. On the seat was a pile of warm coats, so, feeling tired, cold, unloved and unwanted, he curled up under the coats and was soon asleep.

Sometime later, Dianne and her husband Roley broke off their tour of the show to come back to the vehicle for coats. Dianne put her hand in to grab the coats — and they moved.

Coming from Mossman, the home of the giant pythons, she immediately withdrew yelling, "Snake!"

Bravely, Roley advanced to deal with it and rescue the coats when Lee's head emerged. Almost simultaneously the put-upon secretary arrived. Lee had a marvellous night with Dianne and Roley shouting him rides on all the things his mother had forbidden and he was delivered back none the worse for wear .

Bill Rea visited the Brafords at Daintree and Crocodile placing his 'C' on about fifty cows and two bulls and, best of all, legitimised the show poddies. We were saved.

I took Mr. Rea back to his plane and returned to Julatten. We had a few days to put in before taking the cattle down the range to Mossman show. There was always a bed for us and a paddock for our cattle at Edmonds' so there we stayed, filling in the time when we weren't talking and drinking tea, by catching up on Lee's school and going to town for annual visits to doctors, dentists, accountants and the like. The day before Mossman, Bill returned to take us and the cattle down and I left Alice in charge while I went out to Daintree with him to muster cattle from there for the prime cattle section.

We did well in the beef section and held our own with the led-ins. At twenty-nine months, Bully was too old to go in the Bull Under Two Years for a third time so we put Cassandra's 'twin' Bismarck in and he took the honours. With three lots of 'led' cattle entered, Toby's, Bill Edmonds' and ours, it was decided to lead the ribbon-winners in the Mossman Grand parade. A first. And rather interesting seeing that most of the parade was normally new cars and farm machinery — plus a few horses.

Our biggest triumph though, was not in the cattle section, but in the equestrian events. Alice, on a borrowed horse, won the much-coveted Lady Rider. Among other seasoned and talented competitors, she beat the Innisfail Amazon.

10

Always the Changes

In case any present-day show circuit follower is interested, our four-show junket cost a little over two hundred dollars plus Alice's wages. Three nights at the Malanda cabins was under ten dollars, our freight bill under fifty, stockfeed sixty-four dollars, the rest our fuel, tucker, nominations and that all-inclusive word 'expenses'.

The old Boss's racehorse, Zarook, bought as a broken-down gelding and swum back to health by trainer Dinko Pecotich, went his winning way down the coastal racetracks culminating in a win in the Queensland Cup. He'd won the Brisbane Handicap just prior to that at long odds, gladdening the schoolgirl heart of the Boss's granddaughter , Mary Ahlers, who'd outlayed two dollars on him at each start since he made his comeback and had amassed a handy nestegg. Unfortunately his luck ran out when he went interstate and though his increasingly good form led fans to expect the best, he was beaten in the Sydney Cup.

1969 was a very dry year. We were extremely pleased we'd bought the boring plant and put bores down or we'd have been in more trouble. Others wanted water too. Bill and Marie set off with the plant to see what they could do. Unfortunately, while we had plenty of water in the ground, there was very little wind and the mills couldn't keep the stock tanks full. We bought a couple of pumpjacks and took them around, bore to bore, when needed to boost the supply.

Good lucerne hay was available at a reasonable price from the co-op. at Biloela and we fed several tons of this as well as our local hay to selected cattle — breeders with calves. Again bad luck reared its ugly head and in suspicious circumstances we lost two cows we were feeding. One was a daughter of Bill's favourite, Lady. I did a bit of a post mortem and sent stomach contents to the Government vet. who diagnosed poisoning. There was a possibility, he surmised, the wagon used to transport the hay had previously carried chemical in a leaky container. Small comfort.

We also struck a disease we hadn't seen before, something like the familiar Three Day Sickness but lacking some of the usual symptoms and displaying a few of its own. A vet. on route to T.I. called while we had one cow isolated with it and diagnosed Mucosal Disease, prevalent in the U.K. and U.S.A. but almost non-existent here except for isolated cases in N.S.W. and South Australia. As the main symptom besides general malaise was a nasal discharge and ulcers on the mucous membranes of the mouth our thoughts — not voiced to each other but later found to be unanimous — ran to the Threat of all Threats — Foot and Mouth.

We treated the heifer the vet. looked at with the chloromycetin he left us. She recovered speedily and there were no new cases. Thank goodness. The dry weather was worrying enough as it was.

Alice brought a young friend back with her from Mossman. Bobbie was extremely good with animals, especially horses, and interested in all that was going on. We certainly appreciated her company and the help she gave. Despite the fact that most of her time was spent in helping us look after the stock she appeared to enjoy her stay.

By October, things at Monto were grim, too, and Alice decided she'd best go home to lend a hand. We'd scrape by without her with the show season over but were sorry to say goodbye. Nonetheless, we fully understood her wish to go home to help her parents and thought more of her for her concern. Of course we all missed her and probably got a bit slack, too, without Alice to keep us on our toes.

The purchase of Ned and Phyllis Cobb's property was successfully concluded, despite the fact that I turned up at 3 p.m. for a 10 a.m. appointment with the Bank Manager we hoped would finance the deposit. I arrived feeling rather dusty and travelworn after having two punctures and fuel pump trouble on the way down. Another flat tyre would have stymied me completely but the Bank Manager was sympathetic, accepted my apologies, offerred me the use of his washbasin to clean up and later advanced the necessary finance. The rest was up to us.

I didn't really need another Girl Friday but I saw an ad. in the paper — again. I'd had my three you-beaut offsiders and you'll remember how things go in threes. After a couple of false starts Wilma arrived. A pleasant and capable girl who later went on to run a very successful business of her own, she was at that time suffering the traumas of love-sickness. She thought we were on the Bush Pilots mail run frequented by the object of her affections. When she found we didn't even have an airstrip, she left.

I took her into Laura where she got a job straightaway with some bull-tossing contract musterers further north and closer to her beloved.

Meanwhile Bill decided to sell Crocodile and move on. We inspected some places on the basalt near Mount Surprise and he sent me with Jim Boileau to inspect another place near there, Yarramulla, then a waterless block near what is now Undarra. An old friend Jack Withers who used to work on the adjoining Rosella and Spring Creek advised the inspection and provided horses for us to ride.

It was eerie. So silent. With no water there was no life — bar us and our horses — no wallabies, no birds, no nothing. Even the cicadas and other bush insects that normally provide background music were mute. As we rode around the rim of the old crater, Quinkan Cone, it was spookier than ever. Our horses hoofs couldn't even startle a lizard out of hiding.

Jack was sure underground water would be available and dropped pebbles down chinks in the lava-flow walls to prove it. You could clearly hear the 'clink' as the pebble penetrated the surface of the water hidden there. Later, good water supplies were obtained in bores close to the lava flow but dams that had been tried, leaked. Wrong type of earth or something. I wanted Bill to come and look in case he thought it worth the risk and the additional expense but he wasn't that keen.

Springmount, a hundred square mile block not far from Mareeba and once owned by Bill's maternal grandfather, had been on the market for some time. It was closer to the Daintree properties and a N.S.W. family, the Marriotts, were interested in Crocodile. It looked as if it were time to make a move.

The son, Peter Marriott, came with us once when we were looking around Springmount. He had an eye like a Dumpy Level and picked out several dam sites with no effort at all. When we moved down to Springmount he arranged for his friend Steve Grubb (who had Caterpillars) to put in the dams. Steve even did one dam complete with water. While excavating with the bulldozer he struck a soak with enough flow to fill it without hindering his work.

I kept in touch with Janice, Marie and Alice. It was still very dry at Monto but Alice found time to slip up to Mareeba from time to time and George reciprocated by finding pressing business in Monto. In due course they married. Bill and Marie had a son, Robert, often called in to break their journey when passing Crocodile and when we did move to Springmount, Marie stayed with me when her second child, Susan, was born. Of the three, Janice remained fancy-free. Still at Cherokee, she sent me up a young offsider, who was a great help in the shift down from Crocodile. Robyn was pleasant natured, hard working and pretty as a picture.

Johnnie was at home and was doing an agricultural course by correspondence from Lawes Agricultural College at Gatton. It had us all on our toes and reference books were taken from out-of-the-way corners and studied assiduously. Nancy, still at Rocky Grammar made a change from living-in to boarding privately and attending as a day girl. Billy Kid followed John's footsteps and went to St. Teresa's at Abergowrie.

I saw Nancy off on the Sunlander and called at Primac to pick up something Bill needed before driving Kid to Abergowrie. There I had a pleasant surprise in the form of a distant cousin of Bill's, Pat Williams, who was taking his son Cam back too. Cam had been at Abergowrie the previous year so when Pat said he'd take the two boys and Cam offerred to show Kid the ropes I gratefully accepted. The boys transferred Kid's cases to Pat's vehicle and I returned to supervise Laura (now home again) and Lee's lessons.

With Marriotts buying Crocodile and us buying Springmount, I had a hectic time consulting bank managers, accountants and lawyers in between my usual chores. Robyn was handling the show poddies well but Laura and Lee's disorganised schooling had me worried. Normally when kids from correspondence lessons transferred to a school they were ahead but I wanted above all to keep them up to date so they wouldn't prove to be the exception to the rule. I voiced my concern at Springmount to Shirley who had been a teacher and had, incidentally, once taught our solicitor.

"Don't worry," she soothed in what she thought were comforting tones. "A lot of the children at the school come from families where English isn't the first language. They have A and B sections in each grade. They'll just put Laura and Lee in B and work them up."

I wasn't comforted then, but cheered up considerably later when Laura and Lee did very well in their exams and I heard Shirley was having difficulty teaching her own kids by correspondence. Mothers always set high standards (which aren't necessarily always met) for their own children and Shirley didn't have Mrs. Yeatman's discerning advice to remember that, when Mum is trying to educate them at home, 'children usually take after the father's side'. In other words — don't expect too much!

During my absences, Bill, his off-sider Charlie, the two children, Robyn and Peter, now a permanent resident, got on very well. Laura had learnt to make pikelets and kept the smoko tables well supplied in that regard. The show poddies prospered with the exception of a new young bull, named Peter at Laura and Robyn's suggestion. He was one of the first of Bully's calves and did his old man credit in looks but his education wasn't going as smoothly as it should.

Even when we moved to Springmount he didn't change his ways. He was a one-woman bull. I could do anything with him, but he behaved in an obstructive and flighty manner with others. It was flattering, but more than a bit of a nuisance. Should I get home late someone would say, "The show poddies are all fed and rugged..." and someone else would add, "Except Peter. We left his rug for you." And up I'd go, rug him and bed him down for the night. His behaviour was inexplicable.

However, solutions to mysteries, like Truth, have a habit of coming out and many years later while reminiscing with an old friend about the last days at Crocodile I was enlightened to overhear Lee say, "Gee, didn't Peter Bull buck that time when Robyn and Peter put the flank rope on him!"

While I was safely away with solicitors and papers that had to be signed, a mini-rodeo was staged. The heifers didn't react at all and were flops as bull-riding material but Peter Bull excelled. And apparently didn't forget.

Some time later, but long before Lee's revelation, Peter Bull attended his first Atherton Show. We finished for the night and some of the women handlers wanted to look around the pavilions. Toby Swindley's daughter, Brenda, who was helping with their cattle, entered a fruitcake in the cookery section and was anxious to see how it fared. Laura also had her Pumpricot (dried apricot and pumpkin) jam.

The fly in the ointment was that the fireworks were scheduled to begin. I was loathe to leave Peter though he did appear to have settled well. Barry Cornish and his wife also had Brahmans there and Mrs. Cornish, who still had a few chores to do, said she'd wait to see how the cattle re-acted before she joined us. I was anxious, but with her to keep an eye on things, left with the others. My fears seemed to be realised as after I'd listened to exploding skyrocket after exploding skyrocket and the admiring response from the crowd, she arrived somewhat breathless at the pavilion, drew me aside and whispered "Your little bull's upside down".

Panic-stricken I fled through what seemed to be the total population of the Atherton Tableland. Why didn't she untangle him and stand him up?

I didn't get far in the jam-packed crowd and was intercepted at the entrance.

"What did you think I meant?" she asked, an expression of bewilderment creasing her face.

Drat the woman. She put a restraining hand toward me but I tried to pull away.

"Peter," I hissed.

"Oh," she said and a big smile spread over her face. She reached a hand up to my ear. "I meant your earring."

I was wearing tiny Braford bull earrings, the 'in' thing among female Braford breeders at the time. Feeling very ashamed I apologised and we went back to the pavilion, both, I think, much relieved.

Not long before we left Crocodile, the children decided they wanted to invest their savings which had accumulated again. Bill said they were to have no more cattle until they were home from school and could look after them. Johnnie had a growing herd by this.

The solution was Chihuahuas — a breed of dog fartherest from working cattle dogs as possible and thence least likely to cause trouble about the place. The little dogs weren't cheap and a great deal of research went into the venture before they bought a bone and white coloured dog, Jason, from Sydney and Dusky, a black, tan and white bitch from

Brisbane. Their registered kennel names were much more long-winded and unpronounceable and never used except at dog shows when it was necessary to remember what they were in case the stewards called them out.

Jason came first. Dusky was to be mated to a rare blue-coloured imported dog at a rather exorbitant stud fee. We never got a blue pup and some chihuahua fanciers doubted the two parents had even met.

Jason quickly made himself at home. He was an extrovert and very likeable. He came up with his previous show ribbons, his own diamente-studded collar, blanket, dish and a packet of his favourite biscuits. Poor Dusky came up alone in a bare airline pet-pack in a plane that was hours late in arriving. When we picked her up it was obvious she wasn't at all pleased to meet us. Even introducing Jason, all welcoming tailwags and friendliness, didn't help. She snarled and flattened him.

Despite the inauspicious start, Dusky did adjust to life at Crocodile. She and Jason had the run not only of the house and yard but could walk with us what must've been hundreds of chihuahua kilometres to the cattle yards — providing they behaved themselves and didn't attempt, as Dusky did on her initial visit, to savage the show poddies.

When Dusky had her pups, five boys and a tiny girl not much bigger than mice, the children were overjoyed but it looked as if, as chihuahua breeders, we'd been thrown in at the deep end. Dusky developed the canine equivalent to calving paralysis and if help had not been readily at hand in the form of Frank Staunton the Lakefoy vet., we would probably have lost her and the pups. As it was, the diminutive bitch pup didn't survive the fifth day but after we needled Dusky with the calcium solution Justin brought down for us from Lakeland, she was up and about in the blink of an eye and the boy-pups throve.

The calcium borogluconate antidote to calving paralysis established our reps as cattle husbandrymen when a neighbour's milking cow succumbed when we were at Springmount. Stretched flat on the ground on her ribs you wouldn't have given a bent cent for her chances of living but immediately after receiving the calcium she sat up, shook her head, stood up and wanted to walk off. The cure being little short of miraculous, our standing among the milker-owning tobacco farmers soared high.

Bill trucked about half the cattle, the studs, young breeders and some fat, dry cows and saleable bullocks to Springmount and left the rest for Marriotts who purchased additional breeders to re-stock.

Carting the pets almost needed a full-sized truck. Seven chihuahuas with the new pups, the old black cat, Lee's chooks and ducks. We drew the line at Josie, the pet wallaby, and her family. They stayed at their old home and Bill attended to the transport of his dogs and the family's horses.

Springmount homestead was right on the bank of a springfed creek and surrounded by an extensive grove of poinciana, jacaranda and some mangoes. The stockyards were beautifully shaded and pleasantly cool to work in on the hottest day.

Seven chihuahuas with their new pups, the old black cat and Lee's chooks and ducks were all jammed into the Land Rover.

The creek flat rose steeply to a high bank with just enough room to build a big shed with office, storerooms, pens for the poddies and grain and hay storage. At its back wall was our nearest neighbour, Mossy and Behrend Boer and their tobacco farm.

The house itself had been an old tobacco barn — a familiar regional exercise in re-cycling — raised on high blocks, closed in with kitchen, bath, laundry and living rooms underneath and bedrooms and a verandah upstairs.

The stairs had a story. They were where the cat-footed Broncho White was supposed to have fallen to his death. Broncho's ghost was reputed to haunt the stairway but the closest we got to him was his old car abandoned further down the creek bank. Supposedly a 'remittance man' from the Old Country, for years Broncho was a drovers' cook on the route from Coen to Mareeba. He was a smart dresser, reputed to wear a silk cummerbund when he was really togged up, dark, slim and very athletic. Fit him out in half-mast pants, white stockings, a jewelled coat and a flat hat and he'd have made a good toreador.

Broncho had the first transistor radio we'd ever seen. It was talked about for days. We'd gone to where the drovers were at the Block Fence at Crocodile to 'see the cattle through' and met Broncho and his offsider bringing the horses along to camp. He was holding a strange, black, boxlike contraption to his ear. It was the transistor, large and cumbersome by today's standards but amazing then. We listened wide-eyed as he let us listen in turn to the one station available — Radio Australia.

Broncho had the reputation of being Mareeba's 'midnight butcher', a seller of illegal beef. He had a clientele of satisfied customers, led the police a merry chase on his delivery runs and managed to keep clear of disaster — until that fateful moment on the Springmount stairs.

Bill took a tumble on the stairs too, missing his footing as he hurried down carrying Graham Elme's guitar and broke a bone in his foot. No need for long painful rides or for summoning the Aerial Ambulance, we were living in civilisation. In less than an hour he was in the doctor's surgery soon to be returned home in plaster and on crutches.

His plaster didn't slow him down too much. He could mount and ride my mare, one of his old saddlehorses, leaving his crutches at the receiving yard gates and picking them up on his return to the yard with cattle. The crutches also came in handy when he was drafting. He'd stand with one and use the other as a drafting stick until a recalcitrant cow objecting to being poked at, lashed out and split the crutch's' leg'. We replaced it with a cut-down tobacco stringing-stick and cautioned against the use of crutches as drafting canes.

The children, of course, missed their old home and environs but were recompensed with the friends they made on neighbouring farms and at school. A school bus came to Boer's gate and conveyed them to Mutchilba school and later, when Laura reached High School, to Dimbulah.

The telephone was no longer a novelty since Crocodile was linked-up with its loopline but power without having to start or stop a lighting plant was. Town power was the equivalent to Heaven. I gave Marie my faithful 32 V Mixmaster, creator of all sorts of goodies and replaced it with a super-duper 240v one and a toaster to keep it company. We had advanced to a state of luxurious living. Television reception was said to be poor. We decided not to experiment and didn't miss what we'd never had.

The only trouble I had was adjusting to the electric stove. I awoke one night in a panic to the smell of burning flesh, sprang out of bed and raced downstairs where a visitor was sleeping to find I'd merely forgotten to turn off the hotplate under my boiler of corned meat.

A woodstove simply gets the message and goes out should one neglect to put more wood in its firebox, but an electric hotplate, like the babbling brook, goes on and on for ever.

11

Springmount

Springmount is almost in the suburbs of Mareeba, about a half-hour's drive and bitumen almost all the way. It was just on a hundred square miles — many of which could be measured vertically — or, in modern terms, about twenty-five thousand hectares. During Bill's grandfather's ownership it was much larger, but Tinaroo Dam was built near Tolga to bring water from the wetter east coast across the Great Dividing Range to the drier west, and part of the intricate channel system that made this feat possible, cut through old Springmount. It is on the headwaters of the Walsh River, a tributary of the Mitchell which flows into the Gulf of Carpentaria and the first site considered for the dam, the Nellinga Weir, was on Springmount just below the present Collins Weir. Most of the flat country was resumed for irrigation farms, at that time mainly tobacco.

The house block of about four hundred hectares on the irrigable 'downside' of the channel, had a small area of irrigated tropical pasture to which we later added an additional twenty hectares. Johnnie and I became the irrigators mainly by trial and a lot of error but were given a great deal of help by Charlie Nolan of Irrigation and Soil Conservation. Bill found plenty to do repairing yards and fences, erecting new ones and putting in and equipping bores and dams.

At this time there was a push towards growing soya beans and rice, the latter particularly requiring enormous quantities of water. The irrigators who had pastures were concerned that, should there be a shortfall of available water, the end result for them would be calamitous. Some of the earliest pasture irrigators had very low water 'rights', some as low as a tenth of what the Department of Primary Industries recommended at their experimental station, Parada.

Our anxious queries were countered with the reply that we had never been refused extra water — water 'sales'. That was true but we were very uneasy that the day might easily dawn when there was no extra water to be shared out. We'd certainly be up the creek without a paddle — even if it were a dry creek.

A working committee was formed and drew up a submission which fortunately produced results. The water rights that lagged behind were lifted to a uniform level, equal to the best of the private pasture water rights but not as generous as the D.P.I's own supply. Many of the pasture irrigators had Brahman cattle and, 1971 being the Silver Jubilee of the formation of a register for this breed, inspired by our water successes, we took advantage of the cohesive group we'd just initiated to celebrate the occasion. It was too good a chance to miss to promote our breed, especially when we felt the gods of fortune were with us.

It is a moot point what benefited the Far North Queensland cattle industry the most — the entry of the Bos Indicus Brahmans and Zebus, the spread of an introduced stylo known as Townsville Lucerne or the opening of the U.S. market for our meat. Of the trio, the Brahmans are the only ones still with us but in 1971 we had all three and engaged in our celebrations with great gusto.

Poster competitions were organised for Northern school children with the large, eye-catching, varied and well-researched entries exhibited in Mareeba's main street shop and office windows. In collaboration with the Rodeo Association we were to stage a display of Brahman and Droughtmaster cattle on the grounds at rodeo time. As the rodeo attracted huge crowds from all over the state, it was too good a chance to miss. To show the committee's appreciation of the Rodeo's hospitality, it was decided the Northern Stud Breeders would sponsor a candidate in the annual Rodeo Queen competition, one of the highlights of the rodeo.

It wasn't until almost D-day that we realised potential queen candidates were short on the ground and that any who were available had been snapped up eagerly weeks before. After fruitless attempts and discussions we were about to give up when I had a brainwave. Our daughter Nancy was home filling in a year until she went to James Cook Uni in Townsville. She was younger than the usual student age when she'd matriculated and decided to defer her scholarship for a year.

There were all-round congratulations and sighs of relief — until I acquainted Nancy with our decision on her behalf. To say she was not amused would've been a gross understatement. Our relationship was anything but amicable until she reluctantly attended the first function at which all the contestants met, mingled and began their public relations careers. Under the aegis of everybody's friend, Elaine Bird nee Evans, who'd already jumped through the royal hoops, Nancy found she enjoyed the company of the other girls and the socialising. The rodeo princesses, easily identifiable by their satin sashes, acted as ambassadoresses at the rodeo and the social events that preceded it, helping visitors and explaining the complexities.

The candidate question settled, the next problem was a suitably decorated float for the Parade through the main street on the eve of the weekend's events. By this time the committee were attending to other matters much more important and the artistic creation of a suitable backdrop for our rodeo princess was left to us. Ideas of real-live Brahman calves inside a post, rail and leafy crate were abandoned for a safer contrivance of dyed sheets and cardboard cut-outs.

Judging took place at a dinner organised in honour of the candidates and with trepidation we lined up with the others for the parade through Byrnes Street, aglow with carnival lights, glitter and decorations. Apart from going snail-slow, riding the clutch and nearly burning it out, we made it quite safely without bumping into anyone or anything, or spilling Nancy out of the back of the ute onto the bitumen.

What a relief it was over. But it wasn't. The Rodeo President announced the winners leading up to the main event. Nancy's name wasn't among them but we didn't expect it to be. A hush and then the announcement of the Rodeo Queen — Nancy Wallace. She'd won! Fortunately after the previous queen, Robyn, crowned her successor, Elaine took over. The question of a suitable regally-decorated float was raised.

Thank goodness, Nancy could ride and Elaine's mate (and ours) Roysie had a beautiful piebald quarterhorse mare just waiting to lead a parade. The President was easily won over and he and Nancy headed the parade, both mounted, with Nancy on Roysie's gorgeous mare.

One of the many prizes that went with the title was a trip for two to an island resort. Nancy invited Elaine to join her and away they went for a really wonderful break off the North Queensland coast before Nancy began her university studies.

Although life at Crocodile became so much easier after the advent of the road and the benefits that followed, there were still quite long periods when the road forgot the 'all weather' status bestowed on it by the Main Roads Department and went 'out'. Then we had to resort to the non-perishables stored for just such use in the Wet — rice, dried peas, tinned stuff and when the dwindling supply of bread in the freezer dictated, the tin of yeast and the drum of flour. At Springmount, there was no need for this. Town and the shops were but a short drive away.

At that time, cattle sale day in Mareeba was Friday and Friday was the day the Bush came to town. Even if there were no need to buy or sell cattle, there was a good excuse to attend the sale. Everyone would be there. Under the generous shade of Mareeba saleyards' famous fig trees quite a lot of socialising took place.

Jim Boileau was manager of one of Mareeba's stock and station agencies and a more popular man would be hard to find. His office in the main street was the unofficial Cattleman's Club. Any cattleman or woman in town on business or pleasure seemed to gravitate down the street to Jim's office. The billy was constantly on the boil and some of the lengthy discussions the imbibing encouraged, simmered also.

The sixties had been unusually dry with higher than normal deaths in the breeding herds with calving rates down as a result. Fortunately, the "American market' was purchasing whatever cattle could be marketed to keep its hamburger buns full, so prices weren't too bad, but a mystery disease sneaked into the phosphate deficient Peninsula where cattle at times chewed bones to supplement their phosphorous intake when mineral licks weren't available.

As many of these bones were from victims of this new disease, the bone-chewers were also terminally infected and died, apparently inexplicably but none-the-less permanently. Botulism had reared its ugly head and expensive vaccination programs had to be integrated somehow or other into the free-and-easy open-range management of most of the Peninsula,

Another imposition also viewed with scepticism was a levy on stock sales brought in by the Government to finance a Buffalo fly eradication plan. Needless to say, the fly is still with us, though the levy has been replaced by a larger one with a different name.

I became quite expert at calculating the Buffalo fly levy as most of Friday's working hours were taken up by the sale itself and what remained by Jim's open-handed smokos. I usually stayed after the sale to help Jim and his offsiders, Lyn and Kevin, compile the account sale slips for the sale vendors. Jim liked to get the cheques out as soon as possible and it was one way to save him spending weekends at the office catching up.

The amount that came out of the gross sale price — commission, classing and drafting, yard dues (payable to the saleyards) and the accursed levy — to say nothing of the freight charges from home base — was hard to reconcile. It seemed to take so long to produce a sale animal and such a short time to lose a sizeable proportion of that sale price. It hasn't changed over the years, either. The rising cost of necessary inputs are still beating us.

Ted Marsterson, whose son Eddie and his wife Joy, bought Butcher Hill from Bill's family was an ardent advocate of the superiority of the black soil downs over just about any other type of country for fattening cattle. He was particularly interested in the downs country around Hughenden where, after years of droughts and the wool recession, many sheep station owners put their properties on the market. Bill accompanied Ted and Jim Boileau on one trip when they inspected one of the 'dress circle' of Hughenden properties — Hughenden Station — later the home of Ted's younger son, Vic.

Jim also sang the virtues of the fertile black soil downs. Bill was unsettled after leaving Crocodile, part of his old home, Butcher Hill, where he'd spent most of his life. In that restless state he was open to conversion. Springmount was 'small' and civilised by Crocodile standards. Its hundred square miles seemed minuscule after Crocodile's five hundred with the Gresley and Etea blocks.

Soon after arrival at Springmount we lost fat cows and bullocks to heartleaf poisoning. They were ready to market but, not realising the old fences shutting the poison country out were in disrepair, we put them aside for a later sale. They sampled the pretty red-flowered bush with its notched-end leaf and became blood and bone fertiliser. At first the dread botulism was suspected but after Bill rode around the hillslopes and saw the patches of poison bush, it was acquitted of the charge.

Jim began producing enticing particulars of Hughenden properties and the inspection tours began. He was an excellent tour guide, always arrived punctually but contrived also to leave an abundance of time for diversions. If a creek bank were particularly inviting at a hot noonday stop, there seemed to be adequate time to linger and drain the last drop in the tea billy. With his extensive pastoral experience in the Gulf and lower Peninsula he knew the interesting little places hidden away from the general traveller — the lacy waterfall, the weird monolith, the lonely grave — and took pleasure in sharing his familiarity of them.

We usually left Springmount in the afternoon, camping along the road near a creek or convenient dam, and arriving reasonably early the next morning to have a look around. Our camping grounds were easily found. Jim would not roll his swag until he'd found four stakes from which to rig his mosquito net. We didn't bother, contenting ourselves with swatting the odd invader but Jim, whom the mosquitoes found absolutely irresistible, was meticulous in attending to this chore. For years the sight of the little uprights, some no longer quite upright, at the scene of one of our camps brought a nostalgic lump to the throat. Jim's reasons for erecting the net were practical. As a young woman at Milgarra in the Gulf, his mother contracted the mosquito-borne malaria fever and, with the lack of modern day amenities, died of it.

There were many properties for sale. The long drought of the sixties and the concurrent and on-going wool crisis were more than even the best stock managers could survive. The prices, at fire sale levels, were attractive to the cattle people who were then in a slightly better financial position.

Cameron Downs, taken up by Robert Christison as an adjunct to his original holding, Lammermoor, was the first property we inspected. What surprised us at first was the number of buildings that made up the normal improvements on a sheep station. In the Peninsula with its transport difficulties a 'well-improved' station might boast a house of sorts, a shed and quarters. Here, the homestead resembled a small village with its house, manager's and head stockman's quarters, men's quarters, sundry sheds and, of course, a huge woolshed.

With the years, wool-producing grew less labour intensive and many of the buildings abdicated their original purpose and reverted to mere storage sheds. Not so the old manager's quarters at Cameron Downs. It was promoted and changed into a school. Working bees of parents and friends from neighbouring properties converted it into a neat little schoolhouse with a comfortable self-contained flat for the teacher. Cameron Downs donated enough land for the school and playground and, until mains power arrived, provided electricity from the station's power plant. Convinced of the need for a school and impressed by the community's get-up-and-go, the Government provided a teacher and over the years the school has been staffed by a run of exceptional men and women, many of whom stopped longer than the customary one year stint.

Laura was now at High School but Lee could still attend the Cameron Downs School. It was very tempting, and the school committee were ever vigilant to find more pupils to 'keep the numbers up'.

But it didn't come to pass. I flew to Brisbane to arrange finance — successfully, we thought — and a contract was drawn up. Just when everything looked set and we had arranged to bring cattle down, there was a sizeable hiccup in the financial market and one lender withdrew. Reluctantly we had to relinquish the offer but it turned out well. Our cattle could stay on agistment and the manager, through the bush telegraph, heard Barragunda, a small block next-door to, and once part of, Cameron, was coming up for auction.

"It's a little gem," he told us and made special arrangements for us to go over for an unofficial look around.

Our cattle did extremely well on the Mitchell grass downs which pointed up the truth in the old saying that half the 'breeding' is what goes down the throat. We put Springmount on the market and attended, with our mate Jim, the Townsville auction of Barragunda. We missed out. It was passed in to a neighbour at just below the reserve price.

We gave up hope of outbidding the neighbour and headed for home only to spend a couple of days at Cardwell while the Murray Flats on the highway to the north were a sea of water. We even spotted a big barra in the table drain on the side of the bitumen. There had been rain on our way down but we'd managed to keep ahead of it. Now we'd caught up with it on our return trip.

Finally (after Jim assured an old mate of his from bygone days that we were travelling in a 'company' car and not his private vehicle) Jim's friend towed us through the deepest part with his tractor. All in all, the trip was an experience. At that time properties weren't usually sold by auction and it was a bit scary to make a bid of an extra fifty cents an acre and realise after frantic mental arithmetic that you'd just upped the price by eleven and a half thousand dollars. On our trip to Townsville we'd continually checked road conditions, expecting we mightn't make it to the auction and were pleasantly relieved when we did.

Another pleasant surprise awaited us when we arrived back to Jim's office. The neighbour, who, of course, had just been through the same hard times as the vendor, wouldn't come up in price and relinquished his offer. Tempting terms were offered and we became Barragunda's new owners. Johnnie went down to caretake. Springmount was sold — also on terms which we assumed would take care of our commitments — and we shifted to the Downs.

My efforts at making bread no longer reminded Bill of the struggles of an old cow in a boghole.

Which was how, after a few brief years of living a life of affluence, with town power at the flick of a switch, a school bus to a very good little school, tradesmen who called weekly with bread, milk and fresh seafood, I happened to be back to a wood stove (which also provided the hot water), a temperamental thirty-two volt lighting plant, a partyline telephone and a weekly mail — if it didn't rain. The road went out of action after a decent, heavy shower. The gluey blacksoil, as well as being extremely slippery and treacherous, also had a habit of rolling itself like plasticine around and around the tyres until they could no longer turn. Freeing them again, after they'd 'rolled up', was not a chore to take on when attired in your Sunday best.

Laura boarded in town to attend the High School there and after a few trips to and from Cameron school with me, Lee found he could ride a miniature motorised trailbike and take himself to school. There were times the black soil and/or the 'channels' (what we Peninsularites used to call 'creeks') made getting to school impossible. Teacher was prescient and issued 'Wet Packs' of work that could be done at home for such occasions. The kids who lived at Cameron thought this highly discriminatory. They had no valid excuse and their school went on as normal.

Besides the wood stove there was a gas one for quick and convenient cooking. The oven was almost as good as a wood-fired one but I reinforced my vow never to make another loaf of bread as long as I lived. I was going to stick to it, come what may.

Unfortunately, a very short time later I broke my sacred pledge. The Cattle Crunch, said to have been caused by the Oil Crisis, descended upon us. Cattle prices fell to below cost of production while, as usual, input costs rose. The baker shop changed hands and for some reason — the mailman did leave at dawn — the hot bread was taken from the oven and placed in plastic bags containing each property's order.

The new baker didn't make as good a loaf as his predecessor and a few hours sweating in the plastic bag didn't improve the end product. The crunch came at the end of the month when his account came with an extra charge of forty cents for every bag used.

Thankful for the small mercies of the three year break, I let that sacred vow slip my memory, ordered flour and yeast and again took up bread-baking. At least I'd passed the stage when my efforts at making bread reminded Bill of the struggles of an old cow in a boghole.

It was more than twenty years since I'd flown up from Sydney to join Ruth in the Butcher Hill mustering camp. It would be another twenty years before I went back North permanently.

We'd squashed side by side in amongst the khaki burrs and the meat ants behind the Laura pub to watch the corroboree at race time and gaped incredulously at the T.V. as men walked on the moon. We'd carried our babies in front of us on horseback and seen low-slung cars pulling caravans travelling the same route.

The day of the drover had passed. Road trains roared down with the cattle in a fraction of the time. I'd raised five children and saw the oldest of them go their own ways, made lots of lifelong friends and grieved deeply at the loss of the dearest of them.

Things changed. It was a great life, full of fun. I wouldn't have missed it for quids.

Books by

Lennie Wallace

Bow Waves in the Bull Dust

Bitten by the Bull Bug

Nomads of the 19th Century Queensland Goldfields

The Battlers of Butchers Hill

Cape York Peninsula

From Nanango to Cooktown